Astronomers' Universe

For other titles published in this series, go to
www.springer.com/series/6960

Fernando J. Ballesteros

E.T. Talk

How Will We Communicate with Intelligent Life on Other Worlds?

 Springer

Fernando J. Ballesteros
Astronomical Observatory
University of Valencia
Paterna, Valencia
Spain
e-mail: fernando.ballesteros@uv.es

ISBN 978-1-4419-6088-7   e-ISBN 978-1-4419-6089-4
DOI 10.1007/978-1-4419-6089-4
Springer New York Dordrecht Heidelberg London

Library of Congress Control Number: 2010931752

Printed on acid-free paper

Springer is part of Springer Science+Business Media (www.springer.com)

To Herminia, for being the star that illuminates my life.

Acknowledgements

Even when a book is signed by a single author, it is never a single person task. Every book is in fact the outcome of several collective tasks in which a lot of people collaborate. Without their participation, the book would never come about. Therefore I want to express here my gratitude to all those who made this work possible.

First of all, I express my gratitude to Alzira's city hall, Bromera publications, and the University of Valencia for making possible the existence of the European Prize of Popular Science *Estudi General*. I consider this essential, since I deem popular science an obligatory duty for scientists. In a society such as ours, modeled by scientific products, it is indispensable to acquire some background in science, in order to face with rigor and skepticism, fully aware, the challenges (someone of them frightening) of the contemporary world. For this, the task of popularizing science is crucial. Therefore, the existence of such a premium, recompensing popular science, is a great joy for me.

I also want to give my gratitude to the editorial team at Springer, who have always been on the other side of the computer to help me in everything during the writing of the book. I thank Maury Solomon for her patience and understanding.

I wish to express my gratitude to my colleagues (and friends) Alberto Fernández, Amelia Ortiz, Bartolo Luque, Javier Díez, Vicent Martínez, Juli Peretó, Daniel Altschuler, and Eusebio Llácer. Your corrections and suggestions, and our fruitful conversations, were a unique help, and these are in part responsible for the book in its final form. I especially thank Eusebio for all his work and help doing the translation of this book into English, Daniel for the prolog of the book, and Juli for his help in the biological chapters of the book. I also want to thank Pablo de la Cruz, Sofia Fuentes, and Guadalupe Almodovar for putting up with my questions and giving me new ideas when I was lacking inspiration. I appreciate the refreshing presence of Filomeno Sánchez, Mª José Rodríguez,

Federico Medina, and Marina Gómez over so many weekends, an indirect but very useful help.

I would like to especially thank my father and my mother, who always trusted in me and always supported me, even when I said that I wanted to study a subject as strange as physics.

Finally, I want to thank with my whole heart Herminia, my half moon, for her love, comprehension, and patience during the writing process, since she has been the one that has suffered me most. She has been the main engine of my inspiration, and a large part of whatever merit this book has belongs to her.

Foreword

For a long time I have been giving scientific lectures in different countries and on diverse topics, generally related to astronomy and to my work at the Arecibo Observatory. No matter which particular topic I am talking about, the same question always comes up: Have we had any contact at Arecibo with "them"? My negative answer does not satisfy anyone. In fact, the answer either confirms their suspicions that there is a conspiracy afoot by higher authorities not to release information or their intentions to deceive the general public. The reasons for the deception have to do with the idea that, as in the movie *Contact*, the received messages contain important and useful information that will bring great advantage to whoever gets it.

Many of us want to believe that extraterrestrial creatures can talk to us, that perhaps they are even living among us, as UFO fans believe. It would be fascinating if it were true, a more than extraordinary discovery, the answer to an eternal question. There is possibly a deep psychological motive in this desire to know if we are alone in this huge universe, and the need to believe in something beyond our limited world, in space and time.

There is no doubt, then, that this topic brings with it many scientific and philosophical discussions, as well as speculations that, on many occasions, fall into pure pseudoscience because of the lack of a reference framework.

In *E.T. Talk*, Ballesteros provides us with this framework, introducing his arguments with clarity, erudition, and humor. Showing a mastery of ideas and relevant facts, the author briefly tells us the incredible story of our planet, to build the basis on which we consider with whom, with what, and how we could establish any kind of communication, or if that communication would ever even be possible.

Among other considerations, the text presents to us the possibility that life could have originated first in Mars and then later "contaminated" Earth, or as Ballesteros states: "Perhaps, after all, the Martians are us." Perhaps, and to me this would answer many questions.

Daniel Roberto Altschuler Stern

Professor of Physics, Universidad de Puerto Rico and Senior Research Associate, Arecibo Observatory.[1]

---

[1]Former Manager of Arecibo Observatory (1991–2003).

Preface

On July 27, 2010, a public announcement was made in the scientific community concerning the expected ending of a long search: the unequivocal detection of a radio signal coming from an extraterrestrial civilization.

The timing with the vacation period, along with the prudence from the team in charge of the discovery, meant that the news went unnoticed by the general public and the mass media. However, the story, published in the academic journal *Radio Astronomy Journal Letters*, brought about a true revolution among professional astronomers. The article was published under the unassuming title "Radio anomalies of unknown origin in G8 stars" [Osterhagen et al. *R. astr. J. Let. 371, 766–770 (2010)*].

The team, led by Professor Maximilian Osterhagen from the Radio Astronomie Institüt Leuercraff, after a 2-year star-scanning period in the galactic south, detected in the star Tau Ceti in the constellation of the Whale, an anomalous emission of radio waves consisting of a series of pulses separated by intervals of silence. The anomaly sprang from the fact that it was *almost* a cyclical signal. Even though it is usual to find cyclical patterns in nature, the strange periodicity of this emission quickly attracted Osterhagen because the temporal interval between radio pulses varied in a strange way. This fact made the team think that perhaps somebody in the orbit of the source had been able, at some point, to intercept the periodic radio pulses (whose origin was not clear either) and thus hinder the reception of some of these pulses. The team then decided it would try to determine which period was the shortest between pulses, to assess if the rest of the periods were integer multiples of this shortest period.

What they learned was that the periods between pulses were integer multiples of the third part of the shortest. This time of one third of the shortest period had to be, thus, the real period that determined the astronomical phenomenon. The surprise was that not all integer multiples were equally possible. On the contrary, they found the following values in increasing order: 3, 5, 5, 7, 11, 13, 17, 19, 29, 31, 41, 43... These values, taken in pairs, have the peculiarity of being twin primes, which is to say, pairs of prime numbers separated by two units. But it was absolutely impossible that a physical phenomenon could produce a numeric sequence like this. Therefore, it would have been necessary for the "phenomenon" that produced this to know math.

So, we must conclude that Osterhagen's team had found our galactic neighbors – an extraterrestrial civilization – at 12 light years away. There could be no alternative explanation.

Such news should have certainly been the most important news of the millennium, in spite of the calm way in which it was investigated and announced. However, it wasn't. That's because the story above is mere fiction, and, sadly, nothing similar has ever actually happened. But perhaps it helped you to experience the possible effect that such an announcement might have on you. How did you react to the previous paragraph? Which emotions, sensations, or thoughts did you have when you read about the discovery of another civilization among the stars?

In many cases, the answer is undoubtedly incredulity. The possibility of this happening seems so improbable that we tend to be skeptical about it. However, in part, we deem it improbable precisely because of the importance and the consequences that this might have, were it real. Otherwise, if the reader accepted at face value the fictitious announcement, he or she would likely feel an intense emotion. Perhaps such thoughts had occurred as, "This is the kind of news that never reaches the public." You might have felt happy (or scared) over the idea that we are not the only intelligent species in the universe. Possibly you would have felt that this was the most important news you had ever received.

Undoubtedly, if it were true, this would be a very important piece of news and with many consequences in our lives. Why would it be so important? What changes could we experience in our daily lives if we found out that an extraterrestrial civilization existed at hundreds or thousands of light years distant?

To begin with, this would give us a new perspective of things, and a new world view. Our experience teaches us that similar turns have had definite consequences in the past, that these have changed our society, though the change might not have been rapid. In the past, it was believed that Earth was the center of the universe. When Copernicus first postulated that this was not true, that Earth was actually rotating around the Sun (and other subsequent scientific advances, especially the work of Kepler, Brahe, and Galileo, proved that Copernicus was right), our position in the scale of things changed as well as the importance of our world in the grand scheme of things. We then started to think that the Sun was the center of the universe.

From then on, scientific development led to a string of Copernican turns, which separated us more and more from the principal role we thought we were playing in the universe. Later on it turned out that the Sun was not even a special star but only an average star among others. Darwin's work showed that we were not the object of an *ad hoc* special creation by a divinity but an animal closely related to other animal species, the differences coming about because of evolution. Estimations of the positions and movement of the stars indicated that these stars, along with our Sun, seemed to be rotating around a common center (perhaps the true center of the universe), which was actually placed extraordinarily far from us. This group of stars rotating around the supposed "center of the universe" had a disk-like shape, and it was baptized with the name of the Milky Way Galaxy.

We thought that the Milky Way Galaxy was, thus, the universe, until the beginning of the twentieth century, when Edwin Hubble proved that those fuzzy spiral "nebulae" spied through the telescope, and what astronomers had considered small dust clouds inside the Milky Way Galaxy were actually incredibly far away from us, so far that they were outside of our galaxy. These were, in fact, other "universes," composed of thousands of millions of stars (astronomers at first called these immense collections of stars "island universes," until the word "galaxy" was popularly used to refer to them). Our Milky Way Galaxy then started to be considered just our local neighborhood in the immensity of the cosmos.

Each of these revolutions, apart from supposing a change in the scientific paradigm, has had its corresponding social and philosophical consequences. It has changed our way of looking at the world, and it has exerted a great influence on all aspects of our human activity, even in artistic trends. We still pretty much believe that we are the only intelligent creatures in the universe, but we can foresee a new Copernican turn that may change this belief.

If we were to find life forms somewhere in space it would mean a demotion in our "privileged" position that would result in huge consequences; it would mean that life, after all, is something common in the universe and not a rarity. If we were to find another intelligent civilization among the stars, the event would definitely impact our society, and nothing would ever be the same. This would surely answer the question of our being alone in the universe. The mere presence of alien life would reveal that it is possible to survive the risks and dangers of technological development. (The odds are that they would probably be more advanced than we are today.)

Furthermore, if we could establish direct contact and have a dialog with them, suddenly all their advances and knowledge would be in our hands. We would discover the technological limits a civilization can reach, and come to possess a mind-boggling library of scientific or philosophical knowledge, the consequences of which we cannot foresee. Therefore, it would be extremely big news for us to establish contact with an extraterrestrial civilization, and thus many scientists are devoted to the active search of our cosmic neighborhood.

On the other hand, even if we searched for a long time and found nothing, we would not have wasted our time, because in the process we would have generated technology applicable to other aspects of our civilization, learned much more about the universe, and reinforced the belief in the uniqueness of our species, our civilization, and our world, in which we are the sole beings that have knowledge of the universe, and this would mean a huge responsibility for us.

This book deals with the possibilities of actual communication with extraterrestrial civilizations under the light that science can shed. It is not the intention of this book to answer all questions posed but to stimulate the curiosity of readers, so that they themselves can look for the answers to some of these important questions.

The book is divided into three parts, in which we will try to answer, respectively, the following questions: "With whom might we establish this communication?" "What methods will we use to discover extraterrestrial intelligent life?" and "How can we establish communication once we find such life?" – assuming that there is somebody or something out there. In the first chapter we will talk about what contemporary science knows about the origins of life and the possible presence of it in the rest of the universe. The second chapter will be devoted to radio astronomy, including the SETI project, and the means to establish contact. Finally, in the third chapter, we will focus on the problem of which forms of communication and what kind of language could be used to speak with a completely different type of intelligence from ours, alien intelligence. At the end of that chapter there is an epilog that briefly reviews one of the most fascinating topics related with extraterrestrial civilizations: Fermi's paradox.

Contents

Contents

About the Author

Fernando J. Ballesteros earned his Ph.D. in physics from the University of Valencia, Spain, where he is now a practicing researcher and astronomer. He is a long-time popularizer of science in Spain, including acting as coauthor of the radio program "The Sound of Science," in Radio Nacional de España (Spain's national public radio service). He is also the author of the book *Astrobiology, A Bridge Between the Big Bang and Life* (Akal, 2008). He was a member of ESA's space telescope INTEGRAL team and researcher at the Spanish Astrobiology Center (CAB). The present volume is a translation of the Spanish version from Bromera, which was the winner of the European Award "Estudi General" given for popular science.

Part I
With Whom? Finding Life
In The Universe

1. A Place for Life

The belief in the existence of extraterrestrial civilizations starts from the so-called principle of mediocrity. This principle postulates that Earth is a normal planet that rotates around a normal star, which in turn is located in a normal galaxy. That is to say, there is nothing so special in our world as to make it unique. This is a logical conclusion, toward which we are guided by the successive "Copernican turns" that science has suffered throughout its long history, and which has removed us from the central position we once believed to occupy in the universe.

We have come to recognize that both our star, the Sun, and our galaxy, the Milky Way, are typical examples, similar in all ways to those other millions of space objects we have observed with our telescopes, and there seems to be nothing special in them. All this leads us to think that our planet and our Solar System must also be typical examples of the planetary fauna, though knowledge of other solar systems (containing the so-called extrasolar planets or exoplanets) has just begun to be acquired. If this is true, if our world is an common example in the universe, there should logically exist a good quantity of inhabited planets, a fraction of which will contain intelligent beings and civilizations. This is the basic argument to back up the work of all the scientists who actively search for signs of extraterrestrial civilizations.

The majority of the scientific community agrees with the principle of mediocrity, for whenever we have believed ours was a special case, we have painfully come to discover that we were wrong. So it seems a useful guide. But are Earth and the Solar System actually representative cases?

F.J. Ballesteros, *E.T. Talk*; Astronomers' Universe,
DOI 10.1007/978-1-4419-6089-4_1,
© Springer Science+Business Media, LLC 2010

A "Normal" Solar System

According to what scientists know today, the rocky worlds, such as planets or giant natural satellites, constitute an indispensable link in the string of cosmic events leading to life. These bodies are big and stable enough for the chemical elements present to interact in high concentrations, resulting in a variety of interesting chemical reactions. Therefore, if we can learn how common these worlds are in the universe, we can also better measure the possibility of life in other corners of the cosmos. But to make an estimation of something unknown with a reasonable chance of success, we must start from the cases we already know about.

What do scientists know today concerning the origin of our Solar System? A long, long time ago, in a dark corner of our galaxy, there was a huge mass made up of dust and gas, an immense cloud, which actually was a fragment of a much bigger cloud, a nebula, that contained the mass of hundreds of thousands of suns. Its temperature was about –260°C, only a few degrees above the lowest temperature possible. It was mainly composed of hydrogen and helium, and a miniscule quantity of dust and soot. But its density was so low that in a cubic centimeter there were barely a thousand particles. For us, this practically constitutes a vacuum. In comparison, the air we breathe on a daily basis contains almost 27 trillion[1] molecules per cm^3.

The cloud was very similar to the great nebula of Orion, the muse of so many astrophotography fans. There are many similar ones in our own galaxy, but the gigantic nebula we are talking about here does not exist any longer. It disappeared some 5,000 million years ago, completely consumed in the birth of thousands and thousands of stars. One of those stars was our Sun, formed by one of the minor sections of the nebula, one out of the 200,000 million stars in our galaxy.

Although at the beginning of their birth these sister stars were located near each other, as we see in the close cumulus of the stars making up the Pleiades, today they wander along and across the galaxy, due to the galactic tidal forces that totally disaggregated the cumulus.

---

[1] 27×10^{18}

Regrettably, today it is hardly possible to know which of all the stars we see are the sisters of our Sun.

Hence, when we observe the great nebula of Orion or the cumulus of the Pleiades, we are watching the replay of a process similar to the formation of our own Solar System. But how was the change produced from the previously described panorama? How does a beautiful and cold fragment of nebula, a mass of gas and interstellar dust residing in a dark void within the galaxy, become a brilliant star surrounded by planets?

The answer is gravity, the great motor of any change in the history of the universe. Were it not for gravity, these gigantic interstellar clouds would still be only clouds, which later would disaggregate to the end, forming only a remnant gas layer uniformly covering the galaxy, as occurs with a puff of smoke released into a large room. Nevertheless, gravity made this fragment of the nebula start to collapse in on itself, thanks to its own mass.

Once the gravitational collapse phase started, there was no going back in the evolution of the Solar System. Step by step the contracting fragment became spherical. In its central and densest part, the cloud of gas and dust began to rotate, and due to the conservation of angular momentum law, the more it shrank, the higher became the rotational speed. Because of centrifugal force, this central area of the original nebula eventually became a flat disk in which future planets would form.

However, from a great distance there would not have been much that could be seen. The remaining materials covering the region were dense and opaque enough to conceal what was happening inside. Only the heat emission in the form of infrared radiation could escape. At the same time gravity kept on with its work; the center of the cloud continued to shrink, getting more and more dense and increasing in temperature. When it reached 10,000 million degrees, the fire of nuclear fusion started, and a star emerged – the Sun, its light suddenly illuminating the immense disk of gas and dust around it.

The planets of the Solar System began to form later from this circumstellar disk, through the process of gravitational accretion. The dust particles in this disk played a crucial role. They had a denser mass than the gas molecules, and therefore they exerted a greater force of gravity. Attracted to each other, little by little the tiny particles came together to form bigger particles, with a bigger mass and thus stronger gravity, which attracted other particles, generating a chain process. This led to the formation first of small-sized objects, called planetesimals, and later, as these planetesimals united, to the formation of various huge massive spheres, called protoplanets. Finally, these protoplanets accreted the remaining mass of the matter of the disk.

With time, the disk was almost cleared, all of its material transformed into a series of planets and their satellites rotating around the Sun. Far from being an easy process, this was very violent. When the primitive planets, still very hot, attracted such wastes adrift in the Solar System, these did not gently land on the surface of the planet but impacted violently into it. Earth's Moon seems to have originated as a result of the collision of a gigantic object with Earth. In fact, this period in which the Solar System had just been configured is commonly known as the Great Bombardment. It ended some 3,800 million years ago, and most of the craters we find on Solar System bodies today come from that period (Fig. 1.1).

Fig. 1.1   Artistic image depicting the formation of a planetary system. We can observe a star and its planets already formed with the protoplanetary disk still insinuated. © David A. Hardy/astroart.org/PPARC

Other Worlds, But Far Away

Until recently, all this was a theory, though a very well supported theory with much evidence to back it up. All of the planets of the Solar System are located in the same plane (varying only by a few degrees), and they rotate around the Sun in the same direction (called direct), facts not easy to explain if the planets of the Solar System were not all formed at the same time in a disk rotating around the Sun. But this theory has also been tested in the observational field, for we have recently had the chance to photograph other planetary systems at different moments of their formation. The Hubble Space Telescope has taken detailed images of diverse stellar systems forming, with a dark disk of stardust and gas rotating around a recently born star, true snapshots of our remote past. A good part of these have been observed in the close and huge nebula of Orion, a very active breeding place of stars. In some cases, the disks seem to have vacuum spaces, just what we can expect if these stars have giant planets, which have removed or swept out the material in their orbit (Fig. 1.2).

Spitzer, a space telescope studying infrared radiation, has also been observing the Orion Nebula, obtaining an image in the infrared showing almost 2,300 examples of planetary disk formation around stars! From these data we can estimate that around 70% of the stars in the Orion Nebula have planetary formation disks, which shows that the process that formed the Solar System is very common.

But not only have we observed planetary systems in formation; an enormous number of already formed planets have been found as well, orbiting other stars. The first one was found in 1995, and it was the first direct proof we had that our Sun was not the only star that had planets.

Nowadays, thanks to the improvement of astronomical instrumentation, more than 400 extrasolar planets have been found, and this number increases almost daily. Mostly, these new exoplanets are giant planets (in many cases, with sizes much greater than Jupiter), with small orbital periods and short-period eccentric orbits, very close to the central star, which seems to indicate that these are very young planetary systems. But this does not mean that this is necessarily the rule. Such planets have been found because, thanks

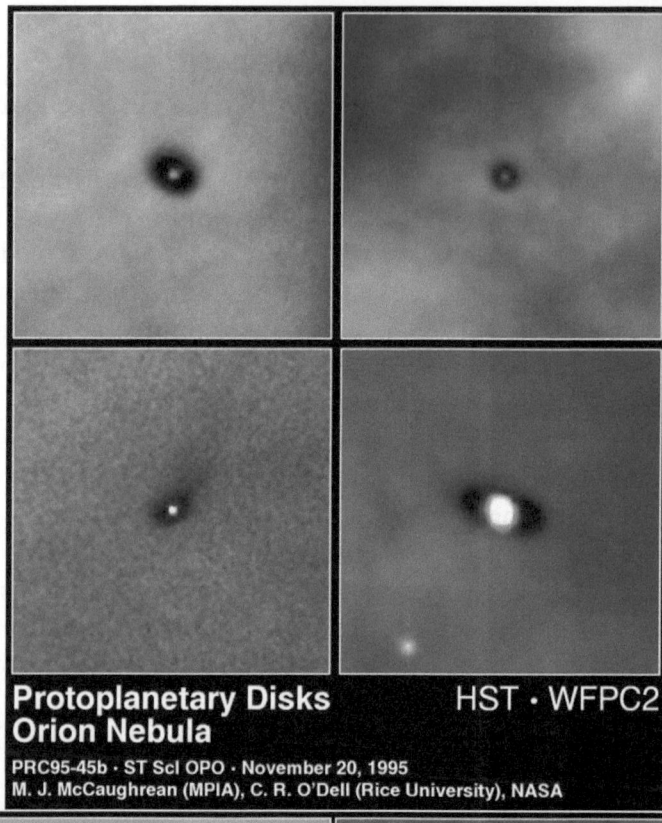

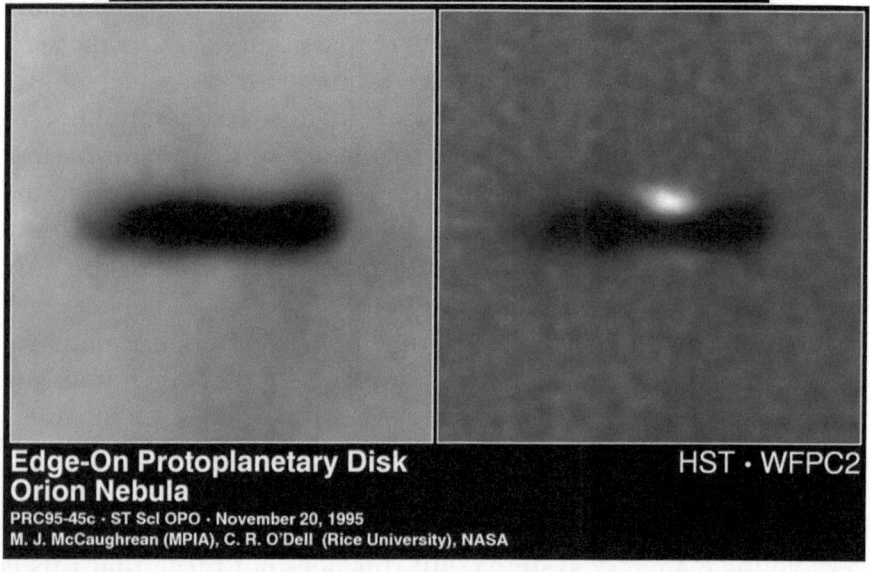

Fig. 1.2 Images taken by the Hubble Space Telescope in the region of Orion showing dust disks around the stars. Courtesy of the Hubble Space Telescope – NASA/ESA

to these characteristics, they are the most evident and easy to find. In addition, many of them have been found in binary stellar systems. This is a big surprise because, for a long time, it was believed that the stellar systems made up of two or more stars could not have planets, as all the material would have been consumed in the formation of those stars. This discovery vastly extended the spectrum of stars that can have planets (Fig. 1.3).

It is expected that with the advance of technology and the new space missions, the discovered number of exoplanets will quickly increase. Among these missions is the French COROT, a space telescope that was built with the participation of the University of Valencia. COROT measures variations in the light of stars, studying, among other things, several stars that are candidates to have planetary systems. If these stars actually turn out to have planets, and it happens that one of these planets passes in front of the star, covering part of

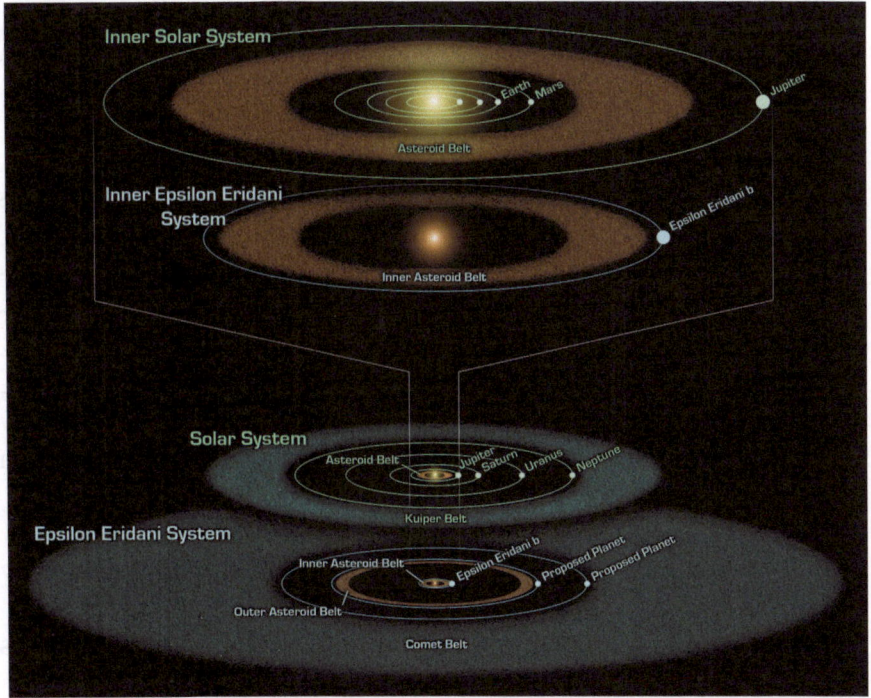

Fig. 1.3 Comparison between the Epsilon Eridani system and our own Solar System. The two systems are structured similarly, and both host asteroids (*brown*), comets (*blue*) and planets (*white dots*). Courtesy of NASA

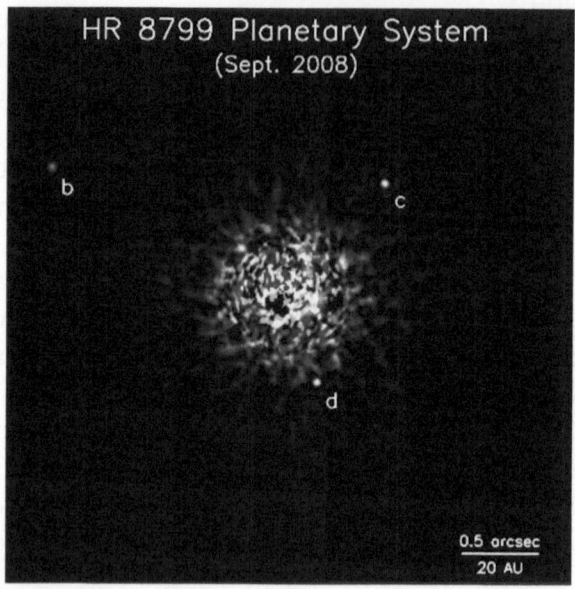

Fig. 1.4   First direct observation of exoplanets in another star, HR 8799, a planetary system with at least three planets, each several times more massive than Jupiter, labeled *b*, *c* and *d* in the picture. The star in the center has been partly eliminated using interferometric techniques to enhance the planets. The image has been obtained using the Gemini North telescope and W.M. Keck Observatory on Hawaii's Mauna Kea. Courtesy Gemini Observatory

its light, COROT will know this by detecting the attenuation of its brightness. So far COROT has found fourteen exoplanets (Fig. 1.4).

Another interesting mission is GAIA, designed by the European Space Agency (ESA) and devoted to measuring with extraordinary precision the position of hundreds of thousands of stars. If some of these stars has a planet orbiting it, its gravity will force the star to undergo a small but detectable swing, picked up by GAIA. With this mission it will be possible to find planets even smaller than Jupiter.

Finally we must mention the Kepler mission, launched by NASA, a complex space telescope mission that was specifically designed to find planets similar to Earth.

But a new technique has been added to the search, and it is proving to be extraordinarily useful: gravitational microlensing. As general relativity shows, the mass of a star deforms the space around it. When a light ray passes near a planet, it undergoes a deflection in its trajectory. This behaves like a lens, and we can use it as such. In ideal conditions, the star can amplify the light, like a magnifying glass, and intensify it. When the light of a much more distant star in the background passes near an unknown planet in its travels towards Earth, the light of the star is suddenly amplified, revealing the presence of the unknown planet. This technique has already detected some extrasolar planets and has proved to be extremely reliable. At present, it holds the record (as published in *Nature* in January 2006) for discovering the smallest extrasolar planet to date – only five terrestrial masses! This is the first confirmed discovery of an Earth-like rocky planet, which is an excellent indication that the Solar System is not a special case.

The Oldest Signs of Life

Knowledge about the formation of our Solar System, and the detection of several extrasolar planets and forming planetary systems, indicate that there are innumerable worlds in the galaxy in which we might find life. But once we have a formed world, how likely is it that life might appear there? Again, to consider this probability, it is necessary to start off from the study of what we know. Unfortunately, very little is known; in this case, all we know about is the appearance of life on Earth. We have only a varied group of theories and some chemical and geological evidence to guide us. Let us begin with geology.

The oldest materials preserved on our planet are zircons found within some rocks of western Australia. Zircon is a very hard mineral that is highly erosion resistant, and for that reason it is common to find zircons older than the rock that contains them. Those of western Australia are thought to be about 4,400 million years old. The interesting thing about them is that they show unequivocal chemical proof that they come from the fusion of a rock previously altered by low-temperature liquid water near the surface. That is to say, these zircons demonstrate that 4,400 million years ago there was already liquid water on Earth's surface and that surface temperatures were not very different from the present ones.

Now let's go to Isua and Akilia in Greenland, where we find interesting old rock formations so well preserved that it is possible to identify their origin. Part of these derive from old submarine volcanic rocks, and other parts have an undeniable marine sedimentary origin. The latter constitutes the oldest set of sedimentary rocks on Earth, since their ages range between 3,850 and 3,760 million years. They are the first direct evidence that some 3,800 million years ago Earth already had oceans with sedimentation on the bottom caused by erosion of old continental crust. This period coincides indeed with the end of the Great Bombardment, the stage in which the Solar System finished forming and the planets were continually being bombarded by rocky fragments. This sounds logical; as long as immense space rocks continued falling on Earth, the energy of the violent shocks would immediately boil off any existing ocean water, becoming steam. Only when the meteoric bombardment finished did it become possible for the planet to have stable oceans (Fig. 1.5).

Fig. 1.5 Sedimentary rocks in Nuvvuagittuq, Canada, older that 4,000 million years. The land around the Hudson bay and the Labrador sea abounds in terrains from the Archean eon. The Nuvvuagittuq supracrustal belt is almost identical to the Isua supracrustal belt but now it seems that they are even older. May be they are the oldest rocks on Earth. Courtesy of NASA

However, these same Greenland rocks are surprising because they show an unequivocal chemical trace of biological activity – an isotopic anomaly in their carbon, a discrepancy between the concentrations of isotopes carbon-12 and carbon-13, analogous to the ones produced by living beings. Not all carbon in nature is the same. This element has, in fact, two stable isotopes: carbon-13 and carbon-12 (in addition to the famous and unstable carbon-14, used in archeology and geology for dating old remains). Although both isotopes can participate in compounds and reactions, living beings will always prefer to use the lightest. This means that organisms and their products are going to be enriched more with carbon-12 than matter not of or from organisms. This is exactly what was found in the sediments of Isua, a greater enrichment of carbon-12.

The biological origin of this isotopic imbalance, published in *Nature* in 1996, has been questioned ever since by diverse researchers. But more recently, in July 2006, a new and more detailed study of these layers seemed to confirm that, in effect, living beings were the cause of this chemical trace, which would mean that life on our planet originated only some hundreds of millions of years after Earth was a ball of fire (Fig. 1.6). In addition if, in the few sediments

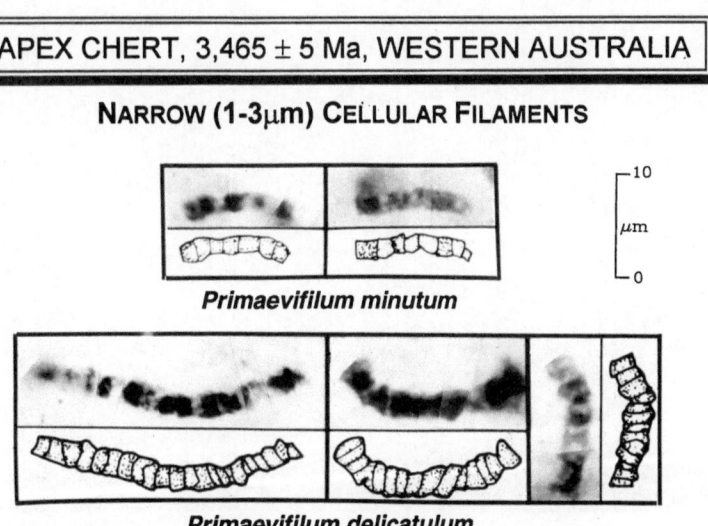

Fig. 1.6   Ancient fossil bacteria from Apex Chert, Western Australia. This fossils, about 3,500 million year old, are among the oldest fossils on Earth. Courtesy of NASA

that we know of from that time, we found tracks of this biosphere from 3,700 to 3,800 million years ago – if this random sample from that past world shows the remains of life – this would mean that life would then have appeared all over the planet.

Other geological data also indicate an early origin of life. Some interesting sedimentary formations, called banded iron forms (or BIF), are marine sedimentary rocks that were formed by alternative millimetric oxide layers of iron and silex, and they are especially abundant in the Archean period and the initial Protero-zoic, becoming notoriously scanter after this. Again, the oldest of them have ages of 3,800 million years, and they can also be found in Greenland deposits. But what do they have to do with life? Well, it is a fact that iron is only soluble in water as Fe^{2+}, so that in the presence of free oxygen in water, iron oxidizes and precipitates. But until life appeared, there could not have been free oxygen in the atmosphere of our planet (Fig. 1.7).

So, how was that iron oxidized? The answer is that we are watching the action of living beings. Iron probably emanating

Fig. 1.7   Banded Iron Formations (BIF's) near Ishpeming, Michigan. Banded Iron Formations are Precam-brian aged, chemical precipitates distinguished by alternating iron-rich and silica-rich layers. Note *lens cap* for scale. Photo by Dr. Sarah Hanson

from submarine chimneys would dissolve in oxygen-free water and would be transported to shallow marine zones, where it would precipitate because this is where the first photosynthetic organisms would be, releasing oxygen into the water.

Finally we move to Pilbara, in western Australia, where we find extraordinary fossil structures called stromatolites. Stromatolites are rocky formations generated by the action of cyanobacteria, in some sense analogous to coral reefs. The remains of bacteria are deposited layer after layer, creating a rock-like structure that grows with time. Today there are still living stromatolites (also in Australia), thanks to which it is possible to recognize in Pilbara what these formations, 3,500 million years old, are. Although at present the biological origin of these fossil structures is widely discussed, and several non-biological alternatives have been offered to explain them, many geologists defend the biological origin of these rocks, so the debate continues (Fig. 1.8).

In any case, all these data firmly support one conclusion: as soon as it was possible to have permanent liquid water oceans (after the Big Bombardment), as soon as the conditions occurred to allow the existence of life, life quickly and easily emerged, in a brief

Fig. 1.8   Presently living stromatolites in Shark Bay, Australia. Picture by Paul Harrison

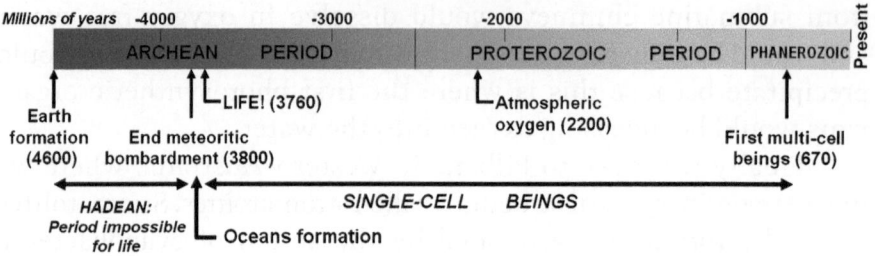

Fig. 1.9   Time line: from the formation of Earth up to the present

time interval (Fig. 1.9). On the other hand, as we can see in the time line of the previous image, complex multi-cellular life took much more time to arrive. Most of the history of the life on Earth was written by unicellular organisms. Does this show the difficulty of taking this step? Maybe this is the bottleneck on the way to intelligence?

2. The Miracle of Life

In the previous chapter, we have seen that it was quite easy for life to appear in our world. But how did it arise? What processes occurred for life to appear so quickly? Are these processes common to other worlds?

The truth is that we do not know much about this. To begin with, we are not even sure of the composition of the primitive atmosphere, which is fundamental to understanding which prebiotic chemical processes led to the formation of living beings. Nowadays the atmospheres of the so-called terrestrial planets (Mercury, Venus, Earth, and Mars) do not have anything to do with the atmospheres they had during the time of the planet's formation. Mostly these are the result of later processes, for example radioactive disintegration of nuclei, evaporation of the ice trapped in planetesimals, volcanic eruptions, or, in the case of Earth, the effect of living beings, which has substantially determined the present atmospheric composition. Therefore, these secondary atmospheres contribute little information on the conditions that reigned on the primordial Earth.

Nevertheless there are some reasons to think that the atmospheres of the giant planets (Jupiter, Saturn, Uranus, and Neptune) are, indeed, close to primordial, remaining quite unaltered ever since these planets were formed, and this can thus be quite a good representation of the original atmosphere that all of the planets of the Solar System (and among them Earth) must have had when they were formed.

In these atmospheres, consisting mostly of hydrogen and helium, carbon is the third most abundant element (even though its abundance diminishes as the orbit of a planet moves closer to Sun). This carbon appears mainly in the upper atmospheric layers of the giant planets, in methane form. Other elements that enrich

F.J. Ballesteros, *E.T. Talk*; Astronomers' Universe, DOI 10.1007/978-1-4419-6089-4_2,
© Springer Science+Business Media, LLC 2010

Fig. 2.1   Harold Urey (1893–1981, *left*) and Stanley Miller (1930–2007, *right*) in the laboratory of the University of California where they conducted the famous 1953 experiment, in which the basic building blocks of life were created. Courtesy of University of California

these upper atmospheres are nitrogen, sulfur, and phosphorus. Using these data as a guide, in 1953 Stanley Miller carried out a famous experiment that moved theory to active experimentation in the study of the origin of life (Fig. 2.1).

Miller suggested to the director of his thesis, Harold Urey, that they simulate the conditions of the primitive Earth in the laboratory, including the conditions of the atmosphere as well as the oceans. For this he devised a system of tubes whereby the gases emulating the atmosphere would mix: methane (CH_4), ammonia (NH_3), water steam (H_2O) and molecular hydrogen (H_2). This was an atmosphere where "reduction" was common. In chemistry, reduction is the term given to the process that is opposite of oxidation. The atmosphere of Earth today, with its elevated molecular oxygen content, is very oxidizing; if we left an iron nail outdoors, after some time it would become completely oxidized. In a reducing atmosphere the opposite happens. If we left an oxidized nail there, it would end up as deoxidized (that is to say, "reduced"), losing all the rest of oxide it contained and staying as new.

Once Miller had the approval of the director of his thesis, he got to work. He set up a tubular system that led to a flask on top, where the gases were mixed; at the bottom, in another flask, was water, meant to simulate the ocean. The whole system was hermetically closed so that the gases could not escape (Fig. 2.2).

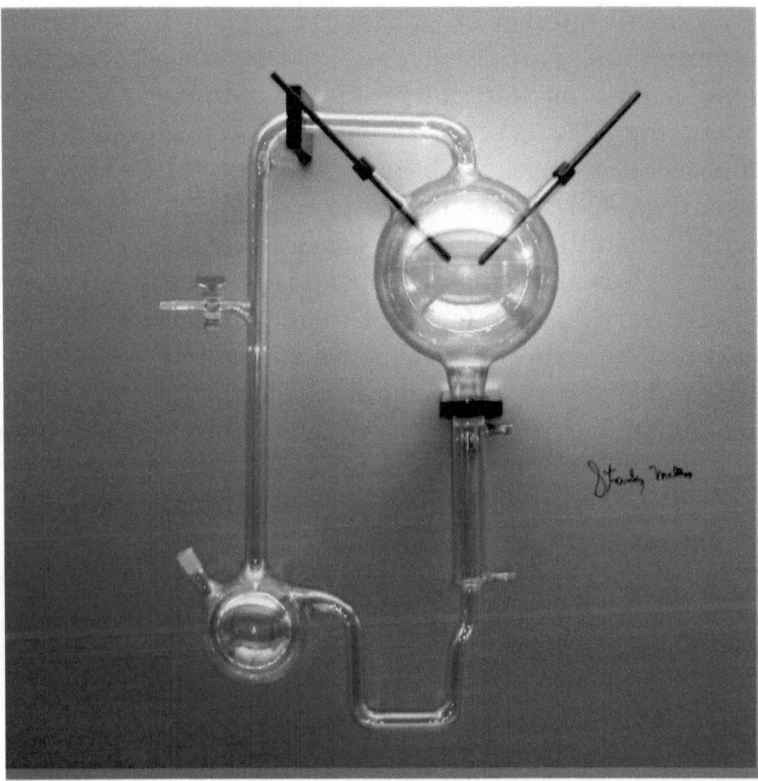

Fig. 2.2   Reproduction of Stanley Miller's experiment, located at the Institut Cavanilles de Biodiversitat of the University of Valencia

In order to emulate the energy sources in the primitive atmosphere (ultraviolet radiation from the Sun, or rays due to storms in the atmosphere), electrodes were used that generated electrical discharges in the top flask (the atmosphere). In order to simulate the cycle of evaporation and later condensation (that is to say, rain), the bottom flask (the ocean) was warmed up so that the water evaporated, making contact with the reducing gases, causing the gas to cool off soon after this in a condenser. In absence of electrical discharges, this rain was clean. But when sparks were produced, something magical happened.

After several weeks of operation, Miller found "sediments" at the bottom of his "ocean." Chemical analyses showed that this sediment was formed by molecules much more complex than the initial ones. The result was not an indiscriminate mixture but

mainly a small number of substances of great biological importance: amino acids (such as glycine, alanine, and aspartic acid) and
urea, molecules that were the precursors of life.

Since then Miller's experiment has been repeated on innumerable occasions, with variations. For example, the power source
might change. Ultraviolet light and cosmic rays were shown to
have an effect similar to the sparks of the original experiment, producing the synthesis of complex organic molecules. Another interesting discovery was the fact that there could not be any oxygen
in the atmosphere part of the experiment, even in small amounts,
because it oxidized the gases and the experiment failed. Finally, a
strongly reducing atmosphere, containing ammonia and methane,
was found to contribute considerably to the formation of prebiotic
products. Otherwise, the rate of organic molecule formation was
much diminished.

In such an atmosphere, a great number of complex organic
molecules would be generated and begin to accumulate in the
primordial oceans. This gave rise to what has been baptized as

Fig. 2.3   Alexander Ivanovich Oparin (1894–1980) was a Russian scientist interested in how life initially
began: "if every living being is generated by a previous living being, how did appeared the first one?" As
early as 1922, he proposed a scientific theory of the origin of life, based on a gradual increase of chemical
complexity through a self-organization process. This process finally would eventually have produced the
first living being. His basic ideas were successfully proved in 1953 by Miller and Urey

the "primordial soup," a term often associated with Aleksandr Oparin who, in 1924 in the Soviet Union, was one of first scientists to rigorously explore ideas on the origin of life (though Oparin never used this term in his writings) (Fig. 2.3). In this primordial soup successive stages of more and more complex prebiotic molecular synthesis would have taken place, which would eventually culminate in the emergence of the first living cells, from which the rest of the biosphere would evolve. If this primordial soup ever formed, it had to have happened shortly after the Great Bombardment (Fig 2.4).

Today this scenario on the formation of primordial soup is not so clear. There are researchers who maintain that the primitive atmosphere of Earth was not reductive but rather neutral. In the beginning the composition of the terrestrial atmosphere had to be mostly formed by internal gases emitted by our planet during the numerous early volcanic eruptions on Earth. These gases were, mainly, water steam, carbon dioxide, and sulfur dioxide. Therefore carbon would mostly be in the form of carbon dioxide rather than methane, so that the primary atmosphere of our planet did not in its beginnings have to be especially reductive. And in these conditions complex molecule production could hardly be sustained. For this reason many scientists believe that the external contribution of organic matter, coming most likely from comets and meteorites, was essential for the first prebiotic stages.

Fig. 2.4   The concept of Primordial Soup, a mixture of complex organic molecules dissolved in a primordial sea, generated by non-biological processes that eventually led to life, although not proved, has become familiar even in the pop culture, as this label proves. Image by James W. Brown

A Polymer in the Chimney

Living molecules are divided into two basic groups: nucleic acids (such as DNA and RNA) and proteins. In all cases we can observe long chains of smaller molecules called monomers, aligned like beads in a necklace. In the case of proteins, these monomers are amino acids, similar to the ones formed in Miller's experiment, and in the case of nucleic acids, they are called nucleotides. Molecules formed by monomer chains are called polymers (to polymerize means basically to form into chains).

In living beings every type of polymer has a different task. DNA is the universal warehouse in which all the information of the living being is stored, universal in the sense that it is shared by all living beings. On the other hand, proteins are the components, the bricks, with which living beings are built; our skin, hair, and nails are formed by proteins. Furthermore, proteins also have the ability to act as catalysts for the chemical reactions of life. RNA, on the other hand, works as an intermediary between DNA and proteins.

The problem is that all of this chemical machinery is closely intertwined and works as a whole: proteins are coded in DNA, and it is DNA that produces them. (When a gene produces a protein it is said to be "expressed.") On the other hand, proteins are what catalyze the replication of DNA. That means proteins cannot be created without DNA, but DNA cannot be duplicated without proteins.

This biochemical version of the old dilemma of "the chicken or the egg" presents a big problem for biologists, because it seems quite unlikely that both types of molecules formed so perfectly adapted to each other simultaneously in the primordial soup. The solution seems to be in that a third molecule – RNA – came to act as a link between them. This concept, going by the name of "RNA world," presupposes that in the initial stages of the origin of life, RNA served a double function that nowadays DNA and proteins do separately.

We know that RNA can store information, since there are viruses without DNA that use an RNA molecule to store their constituent information. On the other hand, it has been proven that a certain type of RNA molecules, the so called ribozymes, also have catalytic properties. Therefore, it does not seem preposterous that the "RNA world" has been, in effect, one of the stages through which Earth passed on its way towards harboring life. The delegation of activities to DNA and proteins, respectively, such as information storage and reaction catalysis, would occur much later in biochemical history.

Even if this theory is true, another problem still remains: how to pass from a monomer collection formed in the primordial soup to a polymer with such a complicated chain as RNA. Polymerization is a determining step; the formation of molecules of between 20 and 100 monomers is necessary, so that a primitive catalysis and replication can take place. Regardless, polymerization, or the creation of these giant molecules, is energetically unfavorable; it is easier to have separate monomers.

In order to overcome this difficulty many scientists believe there must be mineral surfaces to act as catalytic support. The minerals most favored by the scientific community are the clays (such as montmorillonite) and pyrite. These minerals serve as "scaffolding" to guide monomers into merging as polymers. There is some real evidence for this model. Laboratory experiments have demonstrated that substances such as adenosine and guanosine, absorbed in montmorillonite, can give rise to RNA polymers.

All these new ideas about the prebiotic world have been supported somewhat by the discovery some years ago of submarine hydrothermal chimneys and the ecosystems surrounding them. These hydrothermal systems seem to contain all the necessary elements to prove the feasibility of those theoretical primitive worlds: a very reductive local environment, a high concentration of minerals and heat, and evidence of true chemical and biological reactors. These hydrothermal chimneys are located mainly near the underwater ridges, and in them heavy metal emanations take place. These emanations cause a strong chemical imbalance that leads to metallic mineral precipitation, the basis of the chemical energy that many of the presently living microorganisms use today in these environments (Fig. 2.5).

But also found here is the synthesis of organic compounds similar to those in Miller's experiment. In addition, many minerals are formed (pyrites and clays) on whose surfaces polymerizing reactions could take place. One of the great advantages of these

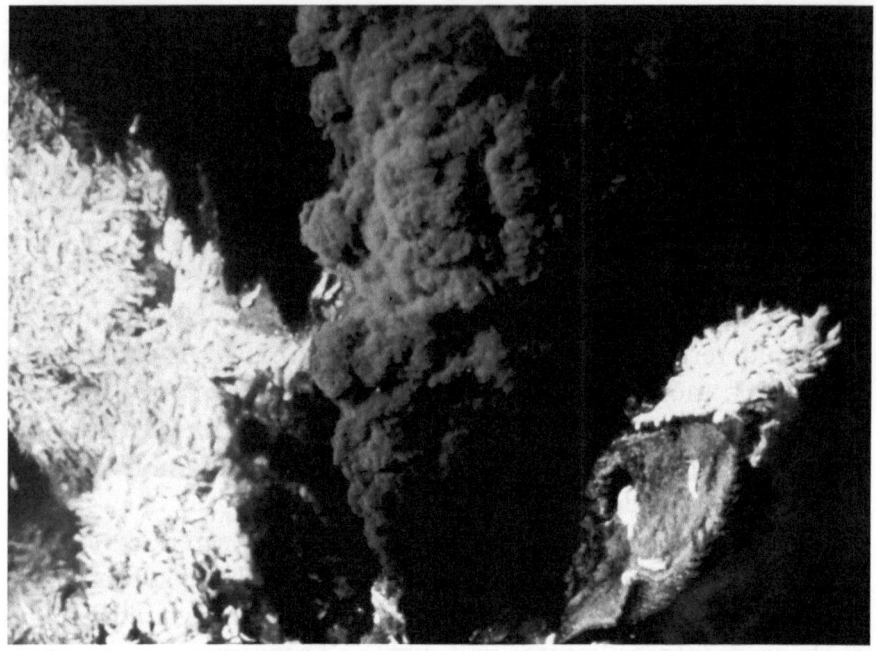

Fig. 2.5   Underwater hydrothermal chimney (Courtesy of the National Oceanic and Atmospheric Administration, or NOAA.)

environments as hypothetical cradles of life is that these processes would not depend on external conditions (atmospheric composition or temperature) for complex organic compounds to be synthesized, which in the end would result in life on Earth.

A Peculiar World?

Without doubt, the ease with which life formed in our world is one of the most powerful arguments to think that another planet with suitable conditions would also give rise to life. But do worlds with "suitable conditions" abound? In other words, is Earth a normal planet?

Some scientists think that, after all, our world is very special; that this is the only planet where intelligence and civilization have emerged. The argument basically says that the emergence of multi-cellular complex life such as is found on our planet (a necessary requirement for intelligence to arise) is extraordinarily improbable, and that only a surprising accumulation of improbable coincidences have allowed this to happen.

The probability that this chain of coincidences will occur may be as low as, let's say, one in a quadrillion (a one followed by 24 zeros). But the universe is very big, perhaps even infinite, or close to it. If in the whole volume of the observable universe there are more than a quadrillion planets, the most elementary statistics say that, perhaps, on one single planet that lucky series of coincidences will occur. Well, that planet is called Earth.

The hypothesis that defends the exceptional nature of our Earth is called Rare Earth. It maintains that our planet is a "spoiled" world, a place where a whole series of improbable lucky circumstances have occurred that have led to complex life. Suffice it to say that if only one of them had not happened, intelligent beings, such as the reader of this book, would not exist.

Among these improbable coincidences is the fact that the orbit of the Sun around the center of the Milky Way Galaxy is practically circular. This causes our Solar System to always be located at the same distance from the galactic nucleus, far from the powerful gamma radiations of the super-massive black hole that lives in its interior. Stars with more eccentric orbits have not been so lucky, and every now and then they stray too near the dangerous central area of the galaxy.

Another coincidence is that Earth is at just the correct distance from the Sun, in a region of the Solar System baptized with the name of the habitability zone. This zone is defined as the region in which the radiation of the star can maintain temperatures on the planetary surface high enough to allow the water to stay liquid (whether it is actually liquid or not it may depend on other factors, such as the atmospheric pressure, the albedo, or the presence of greenhouse gases). If Earth were much farther from the Sun, water would be in ice form, and we would not have seas. If it were closer, the heat of the Sun would be too intense, and liquid water would evaporate. In addition, the zone of habitability changes with the evolution of stars. As time goes by and they get older, stars emit more and more energy, so that the zone of habitability moves outwards. When the Sun was a young star, this area included the planets Venus and Earth. Later, when solar activity increased, the zone of habitability expanded, and Venus was left outside. Currently only the planets Earth and Mars are in that zone. By chance, our planet has had a privileged orbit that has allowed it to always remain within this ideal zone (Fig. 2.6).

The defenders of the Rare Earth hypothesis add other elements to the list of planet singularities, such as having plate tectonics. Of all rocky bodies of the Solar System, only our planet has this peculiar surface-level dynamics that allows, among other things, the renovation of atmospheric CO_2 and with it the existence of a carbon cycle, indispensable for life. Plate tectonics is partly sustained by the high internal heat of our planet, mostly coming from the disintegration of radioactive elements inside the planet, which also maintains the nucleus and the mantle in a fluid state. It is also sustained by the existence of liquid water on the surface, which works like a lubricant between plates (in fact, subduction areas are only found at the bottoms of the oceans). Besides, this fluidity of the planetary interior allows the iron nucleus of the planet to continue rotating like a great dynamo, which generates Earth's magnetic field, a magnetic field that is by far the most powerful among the rocky planets of the Solar System. Earth's magnetic field creates an effective cushion against high-energy charged particles coming from the solar wind, protecting life from the harmful effects of this intense radiation.

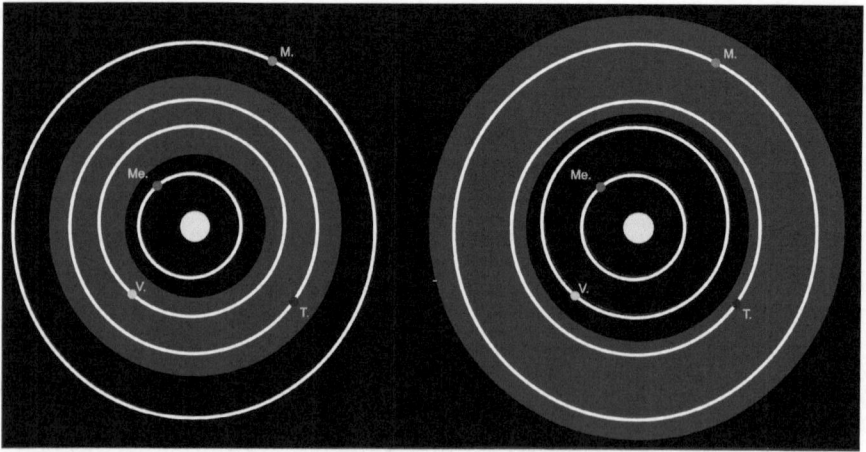

Fig. 2.6 Evolution of the habitability zone in our Solar System as the Sun increased its activity through its life. *On the left*, the young Sun is colder, and the zone of habitability enclosed Venus and Earth. *On the right*, the present Sun, warmer, encloses Earth and Mars

It is necessary to add the existence of the Moon to the list of peculiarities, a giant satellite that stands out in relation to other rocky bodies of the Solar System. Of course, satellites of comparable size exist, and some are even bigger, but all of those satellites orbit the giant gaseous planets. No other rocky planet has a satellite that is so close in size to the body that it orbits. And this is due to the strange origin of the Moon, again a lucky circumstance.

Nowadays there is general agreement among astronomers concerning the origin of the Moon, that it was formed as a result of a fortuitous collision between the primitive proto-Earth and a planetary body similar in size to Mars. As a consequence of this collision, both planetary bodies merged into one, their iron nuclei being united. Remaining in orbit around the resulting planet Earth, though, was a gigantic cloud of matter, a product of the collision, forming a ring. After some time, the ring condensed into a second rocky body in an almost circular orbit; thus the Moon was born. It is important to emphasize that for a celestial body such as the Moon to be formed in a circular orbit, the angle at which the collision occurs must be very precise. Therefore only a single shock out of many could end up producing a double system such as ours. It was completely unlikely to happen, but happen it did (Fig. 2.7).

But how is the possession of a giant satellite related to the origin of life? To begin with, the Moon causes the tides. The life cycles of many coastal species depend on the ebb and flow of the tides. Furthermore, some theories on the appearance of life state that the changes in chemical concentrations brought about by the tides in coastal areas were indispensable for life to emerge on Earth. For example, one of these theories affirms that the prebiotic chemical components that would later constitute the first organic systems indeed formed in those coastal zones, between the tide and the ebb, alternately drying and exposing to solar radiation the chemical compounds, with the subsequent dissolution of these in the sea, where they would chemically react. If these theories are true, and the tides had been less intense, then the chemical machinery leading toward the formation of living creatures would not have been started.

The Moon has also had an influence on the inclination of Earth's rotational axis. On the one hand, the Moon causes the small periodic oscillation called nutation that, along with other orbital

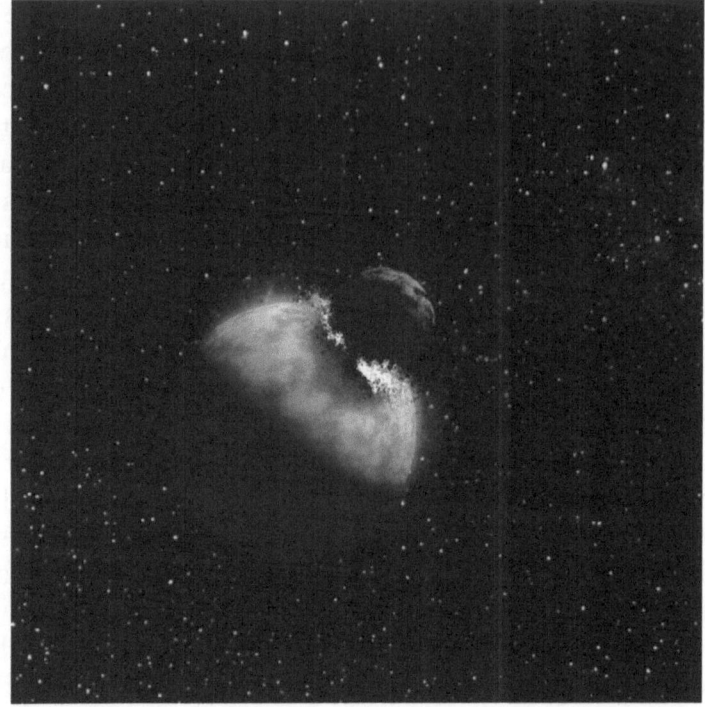

Fig. 2.7 Artistic representation of the impact of the proto-Earth with a planetoid of similar size to the planet Mars, out of which the Moon was formed (Courtesy of the AOES Medialab, ESA 2002.)

effects, produces changes in the insolation received in the different latitudes. It seems clear that these changes of insolation are the causes of the ice ages that have periodically covered Earth with a heavy ice layer and had enormous effects on the evolution of life.

On the other hand, the Moon apparently *prevents* the inclination of the axis from varying. Some scientists believe that the inclination of Mars's axis has undergone enormous oscillations throughout its history, because the gravitational pull of the giant planets Jupiter and Saturn induces chaos in the Martian axis. If Earth did not have the stabilizing effect of the Moon, something similar would happen to it, which no doubt would have had devastating consequences for life.

Finally, we cannot avoid mentioning an ingenious theory by Isaac Asimov, which appears in his novel *Robots and Empire.* The theory is that the lunar gravity prevented the heavy radioactive elements such as uranium from sinking into the planet's interior, at the time when the planet was in a fluid state. As a result, Earth's surface radioactivity is greater than it might have been, thus increasing the mutation rate of living beings, which has led to an evolutionary process that is faster than what would have been possible if the Moon were not present.

However, in spite of the persuasive arguments of the defenders of the Rare Earth hypothesis, for most scientists this position sounds too much like the old geocentrists, who refused to resign themselves to humankind losing its principal role in the universe. In fact, some religious factions and other groups usually welcome this theory, which is in accordance with their religious expectations.

It is not really clear how exceptional the previously mentioned circumstances are; perhaps they are more common than we think. For example, there is proof that in the past there were plate tectonics on Mars. The last data from the *Cassini* probe show that Titan, the giant Saturn satellite, displays signs of fractures in its surface that suggest the existence of plate tectonics. Besides, Io has active volcanism stimulated by the strong tides Jupiter exerts on it, which squeeze the satellite and (due to the heat generated by this friction) help it to maintain its interior in a fluid state, with no need for radioactive disintegration. Thus Io's volcanism plays an analogous role to plates tectonics and renews the planetary surface every few thousand years.

Further, the importance of the habitability zone may be exaggerated. We know that Mars, during the time it was outside the habitability zone, had liquid water in its surface (this is what is known as the Faint Young Sun paradox). Moreover, as we are going to see, nowadays we have information that indicates that Europa, Jupiter's moon, has a liquid water ocean under its ice layer; yet it is totally outside the habitability zone.

In addition, Jupiter and the other giant planets in general stabilize the axes of their satellites far better than the Moon does with Earth's axis. The powerful magnetic field of Jupiter also protects its moons against the high-energy charged particles from the solar wind. Actually, being a satellite of a gaseous giant provides many of the above listed advantages (besides having others, i.e., the giant planet can offer an effective screen against sporadic gamma ray bursts). So perhaps the particular case of Earth, after all, is nothing more than one of many possible scenarios where life can emerge, and it is an error to focus only on considering exact replicas of Earth, as the defenders of the Rare Earth hypothesis have done. Giant satellites around gaseous planets also seem a good alternative, and they abound in the Solar System. In particular, two of them are extremely interesting.

3. Life in the Solar System?

Giant Satellites, Europa and Titan

Europa, the smallest of the four moons of Jupiter discovered by Galileo Galilei, has attracted much speculation on whether it might be inhabited ever since it was observed by the *Voyager* probes. This is mostly due to the fact that water, or better said, water-ice, is the most evident feature of that world, with ice covering this interesting satellite completely, turning it into a smooth white ball.

From the 1980s, it has been postulated that the tidal forces that Jupiter exerts on Europa, although not as powerful as those of Io (Europa is farther from Jupiter), could warm up the interior of this satellite enough to maintain water in a liquid state under the ice layer. Recently, two interesting observations have led scientists to conclude that perhaps this is something more than a theory.

On the one hand, high-resolution images taken from the surface of Europa by the *Galileo* probe show ice formations with some of the same characteristics of Earth icebergs on a frozen sea. It is just as if at some time the dynamics of the satellite had resulted in the ice crust breaking apart, and during a short time the fractured ice blocks had floated, moving and turning, until the underlying water became frozen again (Europa's surface temperature is around –200°C), reforming the ice crust and sealing the fracture (Fig. 3.1).

The second observation is from the *Galileo* space probe, which discovered that Europa has a weak magnetic field that changes its direction according to the satellite's orbits around Jupiter, aligning with the much more powerful magnetic field of Jupiter. But the surprising mobility of Europa's magnetic field can only be explained by the presence of some electrically conductive liquid. The best candidate is none other than salt water.

F.J. Ballesteros, *E.T. Talk*; Astronomers' Universe,
DOI 10.1007/978-1-4419-6089-4_3,

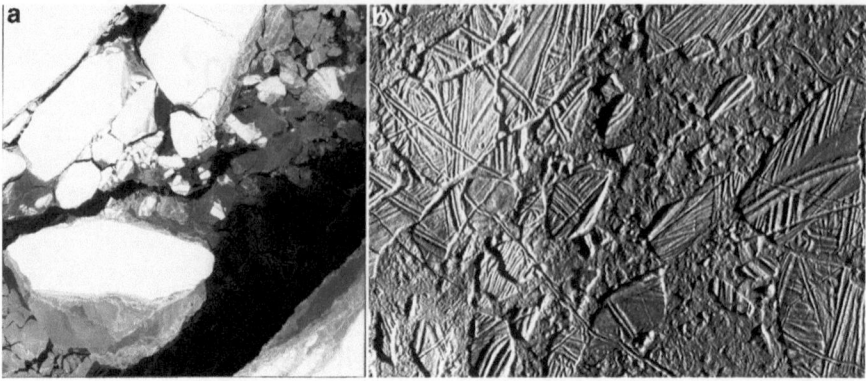

Fig. 3.1 (a) Notice the similarities in Earth icebergs as seen from above and (b) the image by the Galileo probe of the Chaos of Conamara region, in Europa. As we can see, in both cases, the ice fragments can be matched as puzzle pieces. Courtesy of NASA/JPL-Caltech)

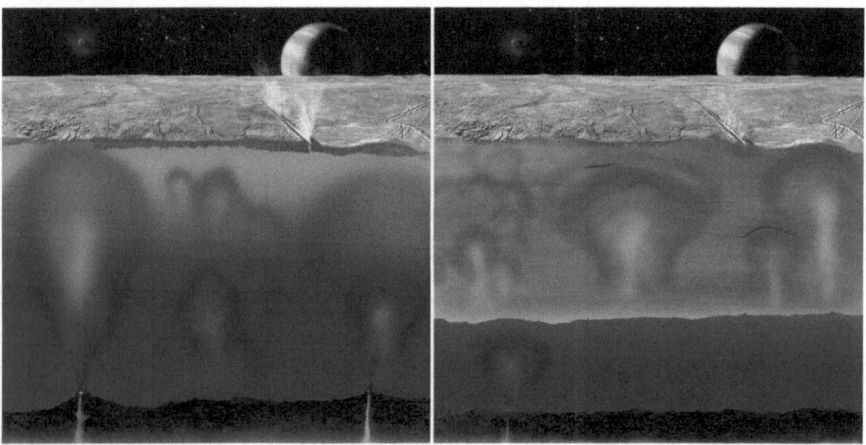

Fig. 3.2 Two possible models for Europa's ocean. Does the moon have a thick or thin ice shell? In both cases, the heat escapes volcanically through hydrothermal chimneys from Europa's rocky mantle. If the heat from below is not intense, the ice shell will be thick (right), but if it is intense, the ice shell will be thin (left) and can break, generating the regions called "chaos", as the Conamara one. Courtesy of NASA

Therefore, both clues indicate that under Europa's frozen crust there exist great accumulations of liquid salt water. One probably would not find small underground salt water lakes; the characteristics displayed by the magnetic field points rather to an almost global underground ocean (Fig. 3.2).

If we accept this theory, then surely the heating from the Jovian tidal forces would also cause an active submarine volcanism, which probably would generate hydrothermal chimneys at the core of the Europan ocean. Hydrothermal chimneys may have played an important role in the origin of life on Earth. We might find that organic life has also been generated in Europan chimneys. Unfortunately, the ambitious NASA mission called JIMO, designed to use a projectile to penetrate the heavy ice layer of Europa and to explore the hypothetical underground ocean, was canceled, so we will have to do with indirect data for now at least to satisfy our curiosity.

Another interesting moon from a biological point of view is Titan, the largest satellite of the planet Saturn. In fact Titan is so big that its size is comparable to that of the planet Mars (Fig. 3.3). Titan, with a surface temperature of –180°C, has a mysterious and dense atmosphere (Fig. 3.4), composed of nitrogen, argon, and methane, a very reductive atmosphere, like the one in Miller's experiment. For this reason it has been said (with more fanfare than proof) that it is an exact replica of the atmosphere Earth had during the period when life emerged, although, as we have seen, it is not known with certainty what kind of atmosphere Earth had at that time. But Titan most likely does have an interesting and rich organic chemistry, whose study could be useful in clarifying some less understood points in terrestrial prebiotic chemistry.

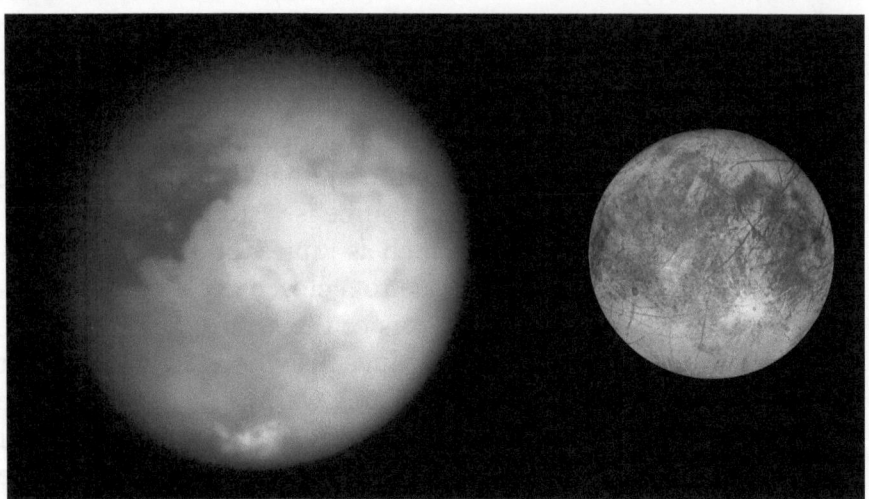

Fig. 3.3   On the left, Saturn's moon Titan (image by Cassini probe), and, on the right, Jupiter's moon Europa (image by Galileo probe). Both moons are shown to scale, according to their relative sizes. Courtesy Cassini/Galileo – NASA)

Fig. 3.4   Josep Comas i Solà (1868–1973, left), Spanish astronomer who first discovered and studied the atmosphere of Titan, Saturn's biggest moon (right, Cassini's picture), from the Fabra Observatory in 1907. Picture of Comas working at the Fabra Observatory, Barcelona, Spain. Courtesy of Fototeca.cat and Cassini/NASA

Part of the mystery of Titan's atmosphere is the presence of atmospheric methane. Certainly there was methane in the primordial nebula that gave rise to our Solar System, and methane has been detected in stellar formation zones. Where is the mystery, then?

The mystery is in the fact that even today we can find methane in the atmosphere, since ultraviolet radiation can easily break down methane (in a process we call photolysis). These broken methane molecules easily recombine with each other to produce more complex hydrocarbons; this way, concentration of atmospheric methane decreases little by little. If we still find methane in the atmosphere today, is because probably another source is restoring it.

In the case of giant planets such as Jupiter or Saturn, the continuous presence of methane is due to a closed cycle of photolysis and regeneration. In the upper layers of the atmosphere, methane breaks down and recombines to produce heavier hydrocarbons that day after day sink down to deeper and warmer layers. As these hydrocarbons get deeper and deeper, the higher environmental temperature decomposes them again into new and lighter methane molecules, which are transported to the high layers, so that the lost methane is recovered.

But this cycle is not possible on rocky worlds such as Earth or Titan. So where does methane come from? In the case of Earth, methane exists due to the action of living beings; it is a product of bacterial digestion. Without life, there would be no methane remaining in the terrestrial atmosphere.

And in Titan? It is almost inevitable to think that in such a reductive atmosphere, exotic organisms have been formed that, as on Earth, are the cause of the atmospheric methane. But with such low surface temperatures it is impossible to have liquid water. Instead, it has been argued that there are lakes or seas of ammonia, or a mix of ammonia and methane that, due to this atmospheric pressure and temperature, can remain liquid on the surface, playing an equivalent role to that of water.

In order to solve the mystery of Titan's methane, a European probe called *Huygens* landed on Titan on January 14, 2005. It traveled along with *Cassini*, and, as it reached Titan, it detached from *Cassini* and collected atmospheric data and images during its descent, until it landed on the surface with a smooth "splotch".

The images taken showed a world with landscapes surprisingly similar to ones on Earth: mountains, valleys, and several strange formations very similar to rivers. The probe found not only atmospheric methane but also evidence of liquid methane on the surface. After the landing, the probe detected an increase of methane levels of about 40%. This revealed the presence of liquid methane mixed with the surface material – the probe had landed over methane mud. Also, the images obtained from space by *Cassini's* radar, even if they do not show oceans, reveal the presence of many surface lakes of some liquid substance (Fig. 3.5).

Nevertheless, if the sole methane source were puddles and superficial lakes, everything would have been lost some hundreds of millions of years ago. The combination of data from *Huygens* and *Cassini* led researchers to conclude that the origin of methane, while not biological, is the consequence of some kind of volcanism that liberates methane from the interior, methane that would have been trapped when this moon was formed.

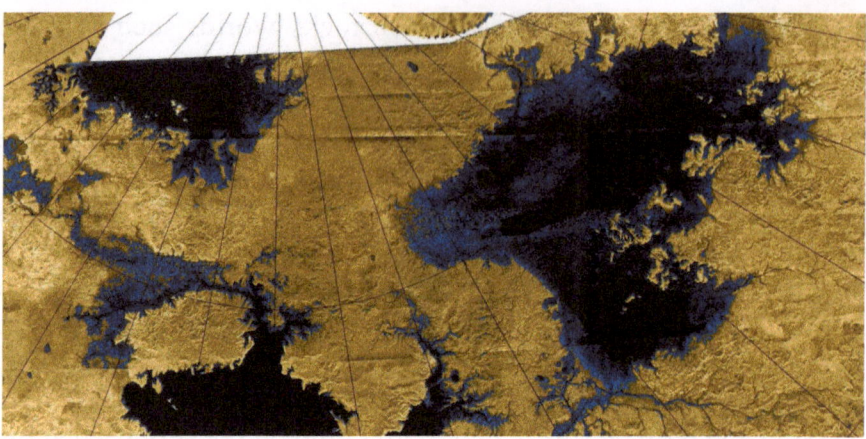

Fig. 3.5   Radar image of Titan's surface close to the north pole, taken from Cassini probe. The picture shows Ligeia Mare (right, measuring approximately 500 km in diameter), Kraken Mare (bottom left) and Punga Mare (top left), some of the ethane and methane great lakes of Titan, besides several rivers flowing into them. Blue areas are liquid ethane/methane, while brown areas are dry land. Courtesy of NASA

Mars, A Watery Past

But if the intent is to look for tracks of life in other worlds, the unquestionable king is Mars, the object of so much astrobiology research. Ever since the controversial Percival Lowell published his book *Mars as the Abode of Life* in 1908, in which he described his fantastic theories on a civilization of Martian diggers of "canals," Mars has excited the imagination of several generations of scientists, science fiction writers, and Hollywood scriptwriters.

Mars mania got to the point that, even among the scientific community, it was taken for granted that Mars was a second Earth and was inhabited. And part of the attraction of our neighboring planet resides in that it does present some surprising similarities with Earth. The length of the Martian solar day is almost the same as ours (24 h 39 min as opposed to 24 h); the inclination of its axis with respect to the plane of its orbit is also very similar to our planet (24°28′ as opposed to 23°27′), which means that it also has seasons; and in addition it displays polar caps, clearly visible from Earth with a simple telescope. Furthermore, both planets have a somewhat comparable size (its diameter is about half of Earth's).

The first high-resolution images obtained from the Mars surface, taken by the *Mariner 4* probe in 1965, were like pouring cold water over our fantastic expectations. These images showed a desert world painfully similar to the Moon. Years later, in 1971, *Mariner 9* went into orbit around Mars, becoming the first artificial satellite of this planet. From this orbit, it drew up a photographic map of the whole planet, the first complete map of its surface, which would serve to help find landing sites for future missions. This map widely corroborated what the images of *Mariner 4* had shown 6 years before. On the surface of the planet, there was neither water nor seas, not to mention canals or Martian cities. As a matter of fact, there was practically no air either, because Mars's atmospheric pressure turns out to be 1% that of Earth's. With such a low atmospheric pressure it is completely impossible to have liquid water, because it boils away quickly and turns to steam.

Nevertheless, *Mariner 9*, as well as the later arriving *Vikings I* and *II* (which arrived at Mars in 1976), found numerous examples of the existence of liquid water on the more ancient surfaces of Mars: multiple dry channels of rivers and structures in form of islands, formed by water flowing. Many of these channels have enormous dimensions, which are only explained by a long history of the presence of liquid surface water. All these spoke to us of a tremendous climatic change that must have been suffered by the planet at some time in its history, causing it to evolve from a world with high atmospheric pressure, rivers, and perhaps seas, and where even life perhaps once emerged, to the dry and cold desert it is now (Fig. 3.6).

Along with the *Viking* and *Mariner* missions, there have been a battery of other space probes that have arrived at Mars in more recent times: the *Global Mars Surveyor* (1997 – NASA), *Mars Odyssey* (2001 – NASA), *Mars Express* (2003 – ESA), the Spirit and Opportunity twin rovers (2004 – NASA) and lately, the *Mars Reconnaissance Orbiter* (2006 – NASA), all of them still in operation at the time of this writing except for the *Global Mars Surveyor*, which stopped emitting data in November of 2006. These missions have shown that Martian geology is one of the most complexes in the Solar System. Mars still has ample regions on its surface that date from the beginnings of the Solar System, the period of the Great Bombardment. Next to these we find modern lands, containing some of the most important examples of geologic activity in the Solar System – the greatest volcanoes, the deepest canyons, and so on (Fig. 3.7).

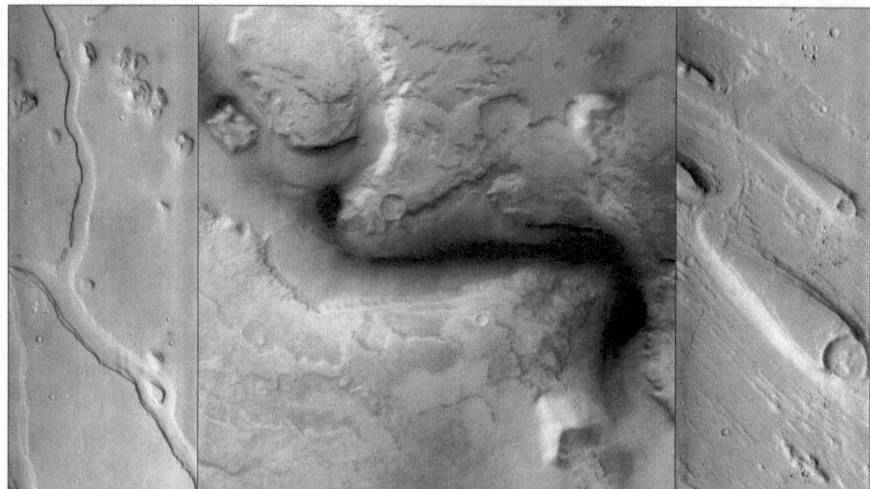

Fig. 3.6 Proofs of ancient liquid water activity on Mars' surface. Left: image by NASA's Mars Odyssey showing a sinuous river bed tributary of Hebrus Valles in the Elysium Planitia region, up to 3 k wide, with a streamlined island indicating that flow was from the lower right to upper left in this region. Center: image by ESA's Mars Express showing an old river bed in Reull Vallis, a region that seems completely carved by water. Inside the bed, old debris deposits can be seen. Right: image by NASA's Mars Odyssey showing streamlined teardrop-shaped islands in the Ares Valles, produced by water flowing towards up left. Courtesy of NASA and ESA

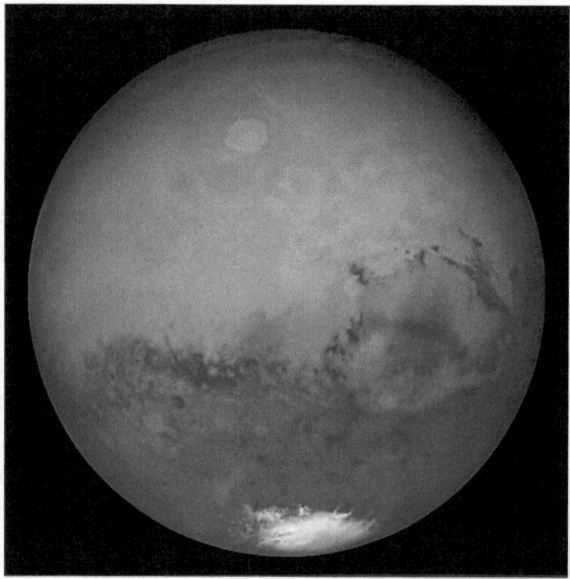

Fig. 3.7 Image of Mars taken from the Hubble Space Telescope. We can clearly appreciate the southern ice cap, and above, in the center, Mount Olympus, the biggest volcano of the Solar System. On the right, close to the edge, we can see the Mariner Valley. Courtesy of Hubble Space Telescope – NASA/ESA)

Surprisingly, the older and the more modern lands do not appear to be found scattered about randomly; they are clearly separated, with the older Martian lands – constituting two-thirds of the Martian surface – appearing mainly in the southern hemisphere, while the remaining third, with the most modern zones, occupying the northern hemisphere. Indeed, in this northern hemisphere, the *Global Mars Surveyor* has found astonishing proof of the existence of an old sea! In that zone the surface is extremely flat, and the border between this surface and the highlands is a horizontal equipotent line. That is to say, it is the coast of an old sea, the Boreal Ocean (Fig. 3.8).

Both Martian rovers, during their more than three years of exploration of Mars's surface, also tell us of a past with abundant liquid water, finding on their travels numerous geologic and chemical evidence that can only be explained by the presence of surface water, not lasting a short period of time but millions of years – sedimentary layers, made up of concretion materials initially dissolved in water that later precipitated on porous rock, and

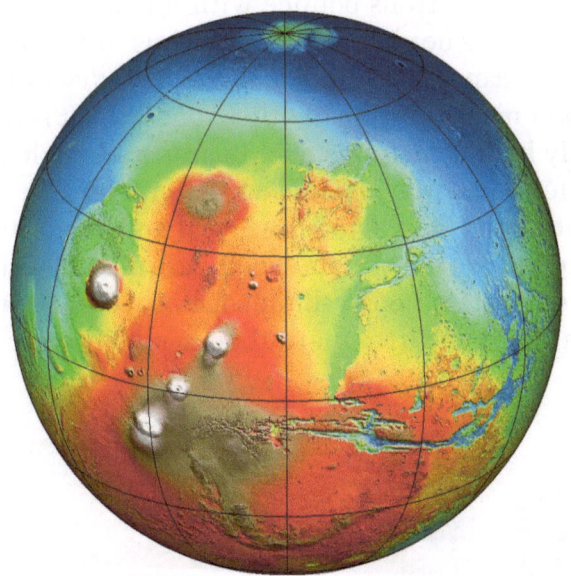

Fig. 3.8 Map of heights of Mars obtained by the Global Mars Surveyor. In the North zone, in blue tones are the lowest lands, where there would have been the Boreal Ocean. Several old rivers are appraised that would end at him. On the left, in white, four mountains (the upper left is Mount Olympus) and on the right, in blue, Mariner Valley, whose interior is at the same height that the Nordic lands of the presumed ocean and that connects with him. Courtesy of MOLA/MGS – NASA

abundant minerals (such as jarosite), which can only be formed in the presence of liquid water.

But where is the water now? Of all the regions on Mars, the only place where there is evidence of water is in the polar caps. Nevertheless, a fast, rudimentary calculation serves to inform us that in the caps there is definitely not enough water to explain all the samples of hydrologic activity found on the planet.

In its search for water, the *Mars Odyssey* uses a neutron detector with which it is able to find water even under the ground. The mechanism used is ingenious and worth a brief explanation. The neutrons this instrument detects come from the Sun. When these neutrons reach the Martian ground, they hit atoms on the ground (although weakly, because the neutron – as its name indicates – does not have an electrical charge). If the shock is against an atomic nucleus of a respectable size, such as silicon, iron, or carbon, the neutron bounces backwards with practically the same energy that it had. But if it hits a hydrogen nucleus composed of only a proton, which has a mass very similar to the one of the neutron, it loses a good part of the energy in the shock, as if they were two billiard balls, so that the neutrons bounce with much less energy than in the first case (or do not bounce at all). Thus, if in an area of Mars there is an abundance of hydrogen, the *Mars Odyssey* detector will measure fewer neutrons than in other zones. What is thus detected are definitely hydrogen atoms, and the most likely hydrogen molecule to be found on Mars is, indeed, water (Fig. 3.9).

The data from *Mars Odyssey* reveal and demonstrate that in latitudes around North 60° and good part of the southern hemisphere there is an immense abundance of ice under the Martian ground to a depth of little more than a meter, with frozen land

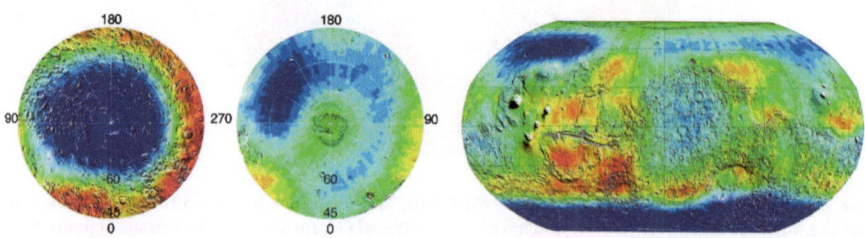

Fig. 3.9 Data from the Mars Odyssey neutron detector. In dark blue, the zones where subterranean water presence was detected. Courtesy of Mars Odyssey – NASA

similar to permafrost. There is indeed more than sufficient water to fill the river basin of the hypothetical Boreal Ocean (Fig. 3.10).

And there are still additional surprises with Mars, because there are traces indicating that part of this underground water could be liquid water! The high-resolution images of the *Mars Global Surveyor* revealed, in 2000, that the walls of craters and canyons have been eroded by water torrents. But what was even more surprising about these gullies was that they are very recent, possibly less than 1,000 years old, since in some cases they had erased sand dunes that, due to winds, are in perpetual movement. These enigmatic gullies are mainly located between South 30° and South 60°, generally coinciding with locations where *Mars Odyssey* had detected ice under the surface.

Fig. 3.10 Artist's conception of Mars, 3,800 million year ago, showing the Boreal Ocean and the Valles Marineris flooded and turned into a sea. The Tharsis region can be seen of the left side of the globe. Courtesy of Daein Ballard

If they are what they seem to be, an explanation might be that these gullies are produced by the sudden emergence of subterranean liquid water at a high pressure. Due to the planet's dynamics, sometimes these subterranean rivers would suddenly appear on the surface, liberating great amounts of liquid water that, although they would boil and freeze at the same time, would still remain liquid long enough to create these torrents, which would last until a new ice cap appears, which seals the water emanation.

If this theory is correct, it is only a question of time until we see the liquid water flowing in one of these torrents. And this may already have happened. In April 2005, when an already photographed Martian torrent was again photographed, it displayed a strange alteration. The torrent, which in the previous image

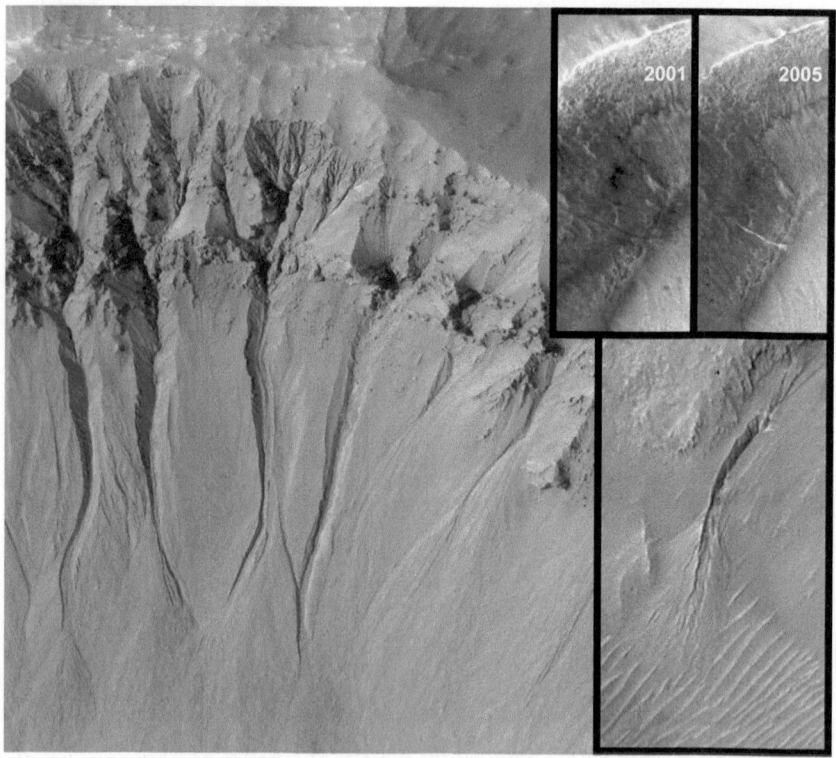

Fig. 3.11   Martian gullies on the side of a crater. Below right, a torrent that has erased part of a series of dunes. Above right, a given torrent observed on two different occasions shows in the image of 2005 what seems to be liquid water flowing. Courtesy of MOC/MGS – NASA

of 2001 appeared dark, in the new image the water seemed to be flowing! If we confirm that it is indeed water and not sand torrents (as another theory maintains), the presence of subterranean liquid water in Mars would bring with it extraordinary biological implications (Fig. 3.11).

Martians!

The more we study Mars, the more surprises we find. In 2004, the spectrometer of *Mars Express* found in the Martian atmosphere methane gas! This discovery, confirmed from Earth by telescopes in Hawaii and Chile, posed a totally unexpected surprise. *Mars Express* found that this methane is not uniformly distributed throughout the atmosphere but is concentrated in certain zones in the intermediate latitudes, zones where torrents have been found. Actually the maps of methane occurrence generated by *Mars Express* perfectly overlap with maps of water created by *Mars Odyssey* (Fig. 3.12).

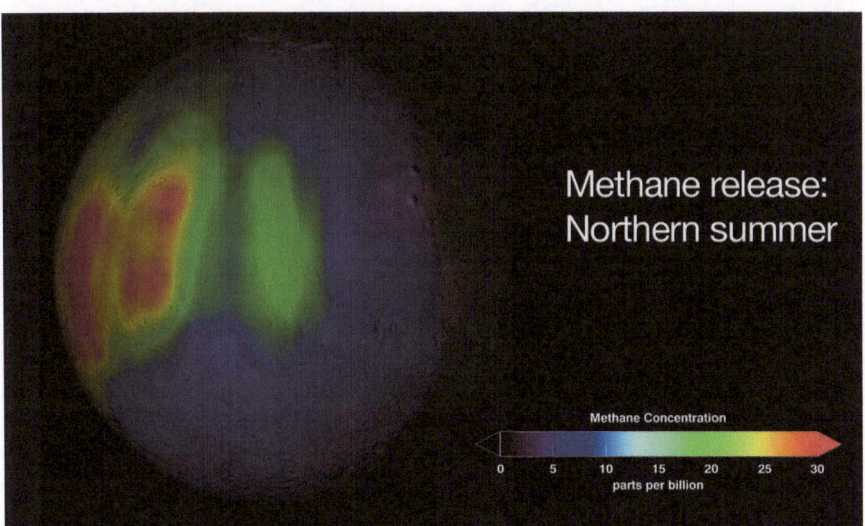

Fig. 3.12 In 2004 ESA's Mars Express detected atmospheric methane which was not uniformly distributed in the atmosphere, but concentrated in some areas, overlapping with the areas where water vapor and underground water ice are also concentrated. This discovery has been later confirmed spectroscopically from Earth using large ground-based telescopes, as this image obtained in 2009 by NASA scientist, that shows an image of Mars superposed with a map of methane concentration. Courtesy of NASA

And again the question arises, where does this methane come from? Since it is such an unstable gas due to ultraviolet radiation, something must be continuously replacing it in the atmosphere. Although the presence of this gas could be explained by volcanism (emanations of trapped underground methane), the fact of not having found any minuscule sulfide trace associated with this methane led us to completely discard any volcanic origin to it. At this time it is extraordinarily difficult to justify the presence of this gas in the Martian atmosphere by means of a geological mechanism. Are we seeing the trace of extraterrestrial biological activity?

The truth is that this is not the first time that we have indications that some form of life can presently exist on Mars. The *Viking* spacecraft that landed on the Martian surface in 1976 had payload experiments on board intended to reveal the presence of organisms in the Martian soil if they were there. To date, these are the only such experiments carried out on Mars. Inexplicably, no Martian rover has the equipment to do biological experiments on board, and

Fig. 3.13 Artistic representation of Viking lander. Courtesy of NASA/JPL

the *Beagle II* probe, which carried on it several different experiments to search for Martian life, ended up crashing into Mars (Fig. 3.13).

The interesting point concerning the *Viking* experiments is that they neither confirmed nor refuted the presence of life in Mars; they produced ambiguous results. The LR experiment of gas release was the one that yielded the most interesting results. This experiment consisted of taking a sample of Martian soil and introducing it into a soup of nutrients the *Viking* had on board. These nutrients were marked with radioactive carbon-14. If in the soil sample there were Martian organisms, these were expected to eat the soup of nutrients and liberate carbon dioxide gas as a product of digestion, which could be detected by the *Viking* sensors.

When the experiment was finished, it was detected that the soup of nutrients was liberating carbon dioxide, a point rather in favor of the possible presence of Martian organisms. But the release might also be due to soil chemistry, so the LR experiment had a second part, which consisted of repeating again the experiment but warming the sample of Martian soil to 200°C before introducing it in the soup of nutrients, in order to kill the possible organisms existing in the sample. This way, if organisms were responsible for the liberation of carbon dioxide, once destroyed there should be no liberation of carbon dioxide in the second part of the experiment. On the other hand, if the gas was a result of the chemistry of the soil, CO_2 should continue being emitted. When this second part of the experiment finished, it was found...

This time carbon dioxide was not emitted! Two points for life. The same result was found by the two *Vikings* located in different areas of the planet and every time the experiment was carried out. Was this proof of Martian life?

In spite of the success of this experiment, NASA officially declared not to have found organisms; they gave two reasons. On one hand, some types of clays, as the above mentioned montmorillonite, can yield similar chemical reactions. On the other hand, another experiment onboard the *Vikings*, intended to look for organic matter on Mars's surface, produced negative results; no organic matter was found in Martian soil. Therefore there could be no organisms.

Later, however, it was discovered that the experiment of detection of organic matter simply was not working! When they later tested it on Antarctic soil, where we know that there are indeed organisms and organic matter, it found nothing. Alas, the experiment was not tested before it was sent to Mars.

Another result of the LR experiment also pointed at a possible organic finding. The *Viking* probes, which were designed to last a few weeks on Mars's soil, continued working for 2 years. During this extensive period the LR experiment continued to be carried out, regularly measuring CO_2 emissions. Surprisingly they saw that the intensity of carbon dioxide emission followed a day/night cycle, decreasing the activity when the temperatures diminished at dusk and increasing with the sunrise. This behavior could not be reasonably explained if the object responsible for the emission was the clay in the soil. It was, however, compatible with the idea that the responsible objects were organisms. What we were seeing could be the result of a biological circadian cycle. Unfortunately, to be sure we will need to return to Mars with new biological experiments.

On the other hand, the excellent quality of the images of the *Mars Global Surveyor* is constantly showing new structures and characteristics of the Martian surface, which have given real headaches to scientists. One example of these structures are black seasonal spots appearing near the polar zones with the arrival of the Martian spring, principally on frozen dunes or inside craters. These dark spots, with a typical size of a few meters, arise on the cap of hoar-frost and grow month after month. They have different characteristics and morphologies. In some places on the planet they present a distinct structure, with a darker central zone and a gray halo around; in fact, these have an aspect that is rather reminiscent of the spots appearing on the skin of mature bananas. Other spots present more exotic forms and look like neuronal networks, or are arranged in the shape of a fan.

Today there are several theories to explain the appearance of these spots, but none is really convincing. Some scientists believe that these are simply thaw zones, regions where the hoar-frost cap has evaporated, uncovering the area below, which seems to be dark in contrast. But this does not explain the form and distribution that these spots present. Another hypothesis postulates that these are originally carbon dioxide eruptions that, with the heat,

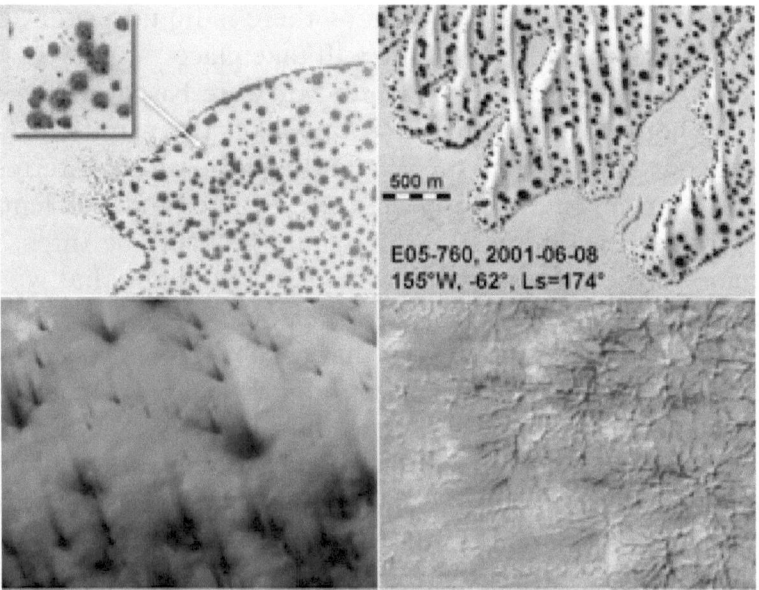

Fig. 3.14   Dark spots appear on the Martian dew at the beginning of spring. Proof of biological activity? Courtesy of MOC/MGS – NASA

sublimate explosively from ice to gas, dragging sand and powder from the soil. On having fallen down, this material would create the dark spots. Again, this theory does not explain why these spots appear principally inside the craters. Thirdly, a group of Hungarian researchers defend an interesting (though not very popular) possibility: that these spots are really colonies of Martian microorganisms, which take advantage of the prosperous time of thaw to reproduce rapidly and then wait to reemerge until next spring (Fig. 3.14).

Life in the Universe

All these traces and indications of presumed biological activity on Mars could end up being vain hopes. Perhaps we are seeing a new version of the "Martian canals" affair, and we are being carried along by our illusions. But maybe not. Perhaps future missions will find life on Mars. And a second case of life in the Solar System no doubt will spectacularly increase the chances that life in the universe is the rule, that as soon as there is liquid water and an energy source, the miracle of life will take place.

What if life on Mars (if we find it) were based on the same organic chemistry, and that it counted on the same DNA and genetic code as life on Earth? Would this be possible? If terrestrial life and Martian life had appeared and evolved independently in both worlds, something like this would probably be impossible. Therefore, the only reasonable conclusion would be that we were related. And this could be so.

Panspermia is an old theory that tried to explain the origin of terrestrial life by a mechanism common in politics – transferring the problem to another place. Life would have arrived on Earth from space. Historically there have been several different variations on this theory, from one that considered that life-seeding spaceships arrived in a remote past to another suggesting that well-formed organisms fell out of the sky, fertilizing Earth (perhaps bacteria, as the British astronomer Fred Hoyle postulated). Still others have suggested that the chemical compounds from which life emerged could not possibly have been created on Earth but fell to the surface after riding on the backs of comets and meteorites. This theory has been completely discredited over the decades, since in fact it hardly explained anything; however, recent discoveries have made scientists re-frame the possibility that life can, after all, fall from the skies.

So why do we think life can fall from the skies? On the one hand, living organisms have been found on Earth that are incredibly resistant to the most adverse conditions. When the lunar probe *Surveyor 3* was sent to the Moon in 1967, it inadvertently carried on board a hundred or so stowaways – the bacteria *Streptococcus mitis* in spore form. Two years later *Apollo 12* landed on the Moon, and the astronauts aboard searched for the television camera of *Surveyor 3*, to bring it back to Earth in sterile condition (just in case it carried possible lunar microorganisms). When the interior of the television camera was opened to be biologically analyzed, no exotic lunar microorganisms were found, but *Streptococcus* spores were, presumably dead. When these traveling spores were introduced into a culture medium, the surprise was huge: they developed into living bacteria that began to multiply (Fig. 3.15). These organisms had survived after more than 2 years under the worst conditions for life – in a vacuum, without nutrients or water,

Fig. 3.15 Astronaut Charles "Pete" Conrad (1930–1999), from the Apollo XII mission (*left*), taking the camera of the Surveyor 3 probe, very close to the Apollo XII lunar module landing site (background). The camera was sealed in a sterile bag and carried to Earth. When it was open on Earth, several spores of streptococcus mitis were found inside the camera that had passed more than 2 years over Moon's surface. Presumably dead, when these spores were put on a culture medium (*right*), they showed to be very alive. Courtesy of NASA

with temperatures oscillating between 150°C above zero to 200°C below zero, and exposed to an intense bombing of solar radiation. Ever since, the list of organisms able to resist the harsh conditions of space has increased. It now includes not only different types of bacteria but also multicellular organisms such as lichens, and even tiny invertebrate animals such as tardigrades.

On the other hand, it is known that material can be exchanged between bodies of the Solar System. The violent impact of a meteorite into a planet or satellite can cause material to be ejected from the planet and, if ejected with sufficient speed, this material will escape the body's gravitational field, becoming a nomad in the Solar System. With time it could even collide with other planets. We know that this can happen because we have already found meteorites on Earth whose chemical and isotopic composition clearly identify them as coming from Mars and the Moon. At the time of the Big Bombardment such an interchange must have been frequent. Is it possible that an ejected fragment from some planet contained living organisms that survived the trip, traveling like stowaways until they fell onto the surface of a new planet?

This is why it is not absolutely preposterous to still think that if life appeared so early on Earth, when there was major meteoritic activity in the Solar System going on, terrestrial organisms could arrive at Mars on board of Earth fragments, fertilizing a fully liquid water world prepared to receive them. The opposite scenario would

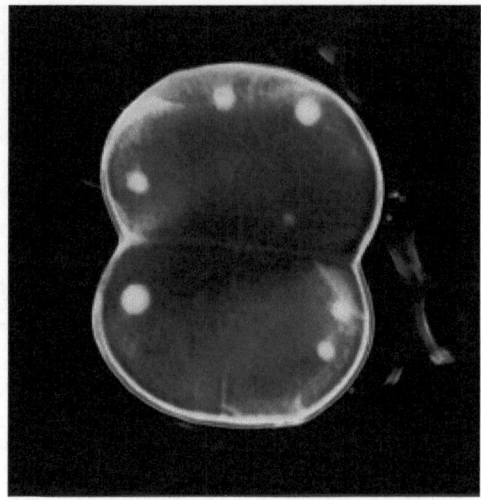

Fig. 3.16 Deinococcus Radiodurans, an extremophile bacteria that tolerates radiation doses lethal for other organisms. (Courtesy of Luis R. Comolli & Cristina E. Siegerist)

also be necessary to consider: that life had originated on another planet of the Solar System (maybe Mars) and arrived on Earth transported by meteorites. Perhaps this explains why life appeared as soon as the Great Bombardment finished. Perhaps, after all, the Martians are us.

To the previous list of organisms that can withstand hostile conditions we can add a peculiar set of unicellular organisms that can not only survive extreme conditions of temperature, pressure, salinity, radioactivity, or acidity but that even thrive in them. Some do well only at temperatures between 80 and 120°C, living in hydrothermal systems. Others live in the ice, at temperatures of –12°C. There are some that avoid humidity and respond well to incredibly dry conditions. All of these organisms, referred to by the name of extremophiles (due to their "liking" for extreme conditions), demonstrate that the conditions of pressure, temperature, atmospheric composition, energy sources, etc., that we are comfortable with actually turns out to be only a very narrow subgroup of the conditions in which life can emerge without problems. The existence of such organisms considerably widens the physical and chemical conditions for a world to be considered habitable, increasing therefore the possibilities that life in the universe is rather common (Fig. 3.16).

Part II
With What? The Search
For Extraterrestrial Intelligence

4. The Search Starts

The previous chapter has shown us that life on Earth arose surprisingly fast and easy, as soon as the conditions were right for liquid water to be sustained on the surface. We saw that there is evidence of the presence of this valuable substance in at least two other bodies in the Solar System, and these are encouraging indications that life in the Solar System might not be limited to Earth. Also, we saw that planetary systems are not rarities of nature but seem to abound across the universe, and that life is much more resistant than was once believed, which extends the limits of what can be considered a habitable world. All these facts, despite our own ignorance about many things, make numerous scientists seem reasonably optimistic about the existence of life in other parts of our galaxy.

Certainly, even if life abounds in the universe, this does not mean it is necessarily *intelligent* life. Maybe there are no other civilizations. But we learned in the introduction the immense benefits that the positive results of this search might bring, making the effort worthwhile even if there is no guarantee of success. Actually, even if we reach the conclusion that we are alone, this would have enormous repercussions in our society. For this reason, hundreds of scientists in the world today are actively looking for any evidence of an existing civilization beyond our Solar System, scientists whose work is often collectively referred to under the name of SETI's initials: Search for Extra Terrestrial Intelligence.

The First Efforts

Early research on extraterrestrial intelligence, at the end of the nineteenth century, was centered for obvious reasons on our Solar System. Certainly, the planet Mars was the principal focus of

F.J. Ballesteros, *E.T. Talk*; Astronomers' Universe,
DOI 10.1007/978-1-4419-6089-4_4,
© Springer Science+Business Media, LLC 2010

these attempts; we might remember that Giovanni Schiaparelli had "discovered" in 1877 the Martian "canals," captivating many of his contemporary colleagues. Among them was Percival Lowell, a wealthy mathematician who loved the science of astronomy.

As we saw in the previous chapter, Lowell believed that Schiaparelli's canals were high-level engineering feats carried out by intelligent Martians, and he dedicated a good part of his fortune to constructing an observatory in Flagstaff, Arizona, to look for signs of extraterrestrial intelligence. Due to this, Lowell was regarded as the founder of the optical SETI (OSETI), which we will deal with later. His active defense of Martian intelligence, through the publication of books and conferences, predisposed the public and many from the scientific community to believe that the Red Planet harbored a civilization (Fig. 4.1).

From early on the use of radio was seriously considered for interplanetary communication, and the pioneers of the Hertzian waves were the ones who made the first attempts to search for extraterrestrial intelligences. Nikola Tesla, who in 1893 constructed the first radio transmitter, detected in 1899 a series of repetitive signals, in coherent groups of one to four clicks, and determined that these were coming from Mars. Here is what he said in his own words: "I have observed electrical actions which have appeared inexplicable, faint and uncertain though they were, and they have given me a deep conviction and foreknowledge that before long all human beings on this globe, as one, will turn their eyes on the firmament above, with feelings of love and reverence, thrilled by the glad news: Brothers! We have a message from another world, unknown and remote. It reads: one... two... three..."

The publication of this news caused Tesla to be discredited by the whole scientific community. But this did not discourage him. Tesla spent the last years of his life trying to communicate actively with his hypothetical Martians. Today it is believed that these signals were natural radio broadcasts from the ionosphere of the planet Jupiter, which are easily detectable by radio. These natural emanations are emitted every now and then in the form of double and triple pulses, which is quite like the description given by Tesla.

Some years later, in 1919, history was repeated. Guglielmo Marconi, inventor of the telegraphy without threads, again detected

Fig. 4.1 Comparison between the Lowellian "observations" of a canal-covered Mars (*left*), and the real images of the same zones, obtained by the Hubble Space Telescope (*right*). Notice that the north is *down* and the south is *up*. Courtesy of NASA

a few strange radio signals, and he determined that these were coming from Mars, which caused a considerable public commotion. Nevertheless, Marconi did not suffer Tesla's discredit, partly because he did not show such a solid conviction as the latter and partly because of the support he received from the renowned scientist Lord William Thomson, baron of Kelvin. Apparently, the signals that Marconi detected were actually Morse code pulses, distorted by the terrestrial ionosphere, from a remote radio station. The ability of the ionosphere of our planet to transmit long distance communication thus began to be discovered.

In 1924, coinciding with Mars in opposition (a time when the planet is closest to Earth), David Todd, a partner of Percival Lowell, coordinated the active scouting of radio signals that could come from Mars, managing to persuade even the American army and the Coast Guard about the use of radio for these purposes. Was he successful? Well, as the opposition passed, on the front pages of newspapers the headlines read: "Radio detection when Mars approaches," and also "Possible Mars flash by radio is repeated," so, in fact, some receivers had heard something. Later the true source of at least one emission was discovered – the received pulses were coming from Seattle!

With the improvement of radio systems, it became evident that no type of radio signal was coming from the planet Mars. At the same time, increasingly discouraging evidence indicated that Mars, the most promising world in the Solar System, was actually a cold, dry, barren desert with a thin atmosphere. This moved the target of the radio scouts beyond our Solar System, an activity made possible with the arrival of radio telescopes.

The Ozma Project

The starting point of the current search for extraterrestrial intelligences with radio telescopes was a scientific article published in 1959 in the journal *Nature* by the physicists Giuseppe Cocconi and Philip Morrison, entitled "Searching for Interstellar Communications." In this article a realistic strategy to look for such intelligences was proposed for the first time, stressing that two radio telescopes of a reasonable size (with parabolic dishes of around 75 m in diameter) should not have any problem communicating with each other even across immense interstellar distances. The article concluded that if such interstellar signals were coming from the nearest stars, we already had the means – radio telescopes – to detect them. A little later, Morrison's and Cocconi's theory was tested (Figs. 4.2 and 4.3).

One year before Frank Drake, a young radio astronomer employed at Green Bank's observatory in West Virginia, had independently reached identical conclusions. After a series of calculations he realized that, of all the radio telescopes that existed in

Fig. 4.2 Giuseppe Cocconi (1914–2008, *left*) and Philip Morrison (1915–2005) published in Nature in 1959 a paper laying the foundations of SETI procedures. It is usually assumed as the beginning of SETI. Courtesy of MIT

Fig. 4.3 The 26-m telescope of Green Bank's observatory in 1960, with which Frank Drake carried out the Ozma Project. Courtesy of NRAO/AUI/NSF and Cosmic Search magazine; http://www.bigear.org

Fig. 4.4   Frank Drake (born 1930), working at the Greenbank observatory when he carried out the Ozma project (main picture, in the *middle*), and at his office at Cornell University (*inset*). Courtesy of NRAO/AUI/NSF and Cosmic Search Magazine (http://www.bigear.org)

Green Bank, one of them was of sufficient size to be able to detect a signal sent by an equivalent telescope located up to a maximum of 12 light years away. And several stars similar to the Sun existed inside this limit. Drake thought that he could do something more than the simple theoretical calculation, and proposed to the observatory management the fantastic idea of using the 26-m radio telescope to look for possible radio signals from other civilizations. To his surprise, he received the authorization (Fig. 4.4).

After 2 years of silent preparations (since Drake was afraid of the criticism his project could provoke among the scientific community), finally, in 1960, he began the search. This pioneering work was dubbed the Ozma Project by Drake, after a character from the Wizard of Oz tales. Since the Ozma Project did not allow for a lot observing time, it was decided to do a long scout of only two stars. These stars had to be similar to the Sun, be approximately 11 light years away, and not be stars in a multiple system. (It was believed at the time that star systems with more than one star could not form planets.) The selected ones for this scout were Epsilon Eridani and Tau Ceti.

Actually, the Ozma Project detected on two occasions a strong pulsing signal when the antenna was pointed at Epsilon Eridani, exactly the kind of signal that Drake was expecting an interstellar communication would have. Unfortunately, the second time the signal was heard, a low-power secondary antenna also detected it, which indicated that the source was really much closer. As a matter of fact, this was a U2 spy plane, which was flying over the zone. As we can see, this type of event has been and still is the general result of these searches. The majority of promising signals detected come from one civilization – ours.

Ozma was the first active search for signals produced by other intelligences in our galaxy, and even though it had negative results, it also demonstrated that a controversial topic such as extraterrestrial intelligence could be approached with scientific rigor. It was the beginning of what we today know as SETI.

The Ozma Project, along with the article by Cocconi and Morrison, provoked an unexpectedly strong scientific reaction, both negative (criticism and scorn) and positive. Suddenly the scientific community became interested in finding other signs of intelligence in the galaxy. The time of little green or pig-headed Martians, an inheritance of the pulp science fiction from the postwar period, had passed. The topic of extraterrestrial civilizations entered its scientific maturity. The proof for this was that only a few years after the Ozma Project, a few mysterious signals that were detected were thought to be extraterrestrial, and there was neither objection nor shame in attributing them to interstellar intelligences.

Pulsing Signals

Some people believe our destiny is written in the stars. The truth is, the destiny of the stars is written... in their mass. Certainly, how a star evolves depends on the quantity of mass it possesses. A star with little mass, or with a medium mass, such as the Sun, is destined to slowly consume its nuclear fuel, its hydrogen (the slower it burns the smaller the amount of mass it has), until towards the end of its life it bloats up and becomes a red giant. However, a little later it will collapse again in on itself and turn into a minuscule dense and warm collection of embers called a white dwarf. This is a relatively calm life, as corresponds to a star with good manners.

Stars that have a higher mass (several times the mass of the Sun) have a different kind of life cycle. Like "rebels without a cause," they live life on the wild side, rapidly consuming their nuclear fuel and, after a brief youth, die in a spectacular form, bursting into some of the most violent explosions known in nature: a supernova.

The intensity of a supernova is such that, during the explosion, it shines more intensely than all the stars together in its galaxy. But the supernova does not cause the star to disappear. In the remains of the explosion there can be found a minuscule body of only around 10 km, consisting only of neutrons so densely packed that a football made of this material would weigh a trillion[1] tons. This curious object is a neutron star. With a magnetic field billions[2] of times larger than our terrestrial one, its magnetic poles generate intense jets of electromagnetic radiation, which are thrown out into space. This is the swan song of a dying star. The position of these poles does not necessarily coincide with the rotational axis of the star, so that these jets usually turn like a lighthouse in the sidereal gloom. If one of these jets happens to point towards Earth, we will detect a signal pulse whenever the star turns. The object is then referred to as a pulsar. Pulsars are characterized by radio pulses emitted with extreme regularity, so regular that they can be used as natural clocks.

The first of these pulses were detected in 1967 by the Ph.D. student Jocelyn Bell. Surprised by the strange appearance of the signal, she showed her discovery to her thesis director, Anthony Hewish. What Bell had seen was an emission of regular radio pulses accurately repeating themselves every 1.33 s. Initially Bell and Hewish theorized that the signals were due to local interference, but soon they discarded theory, since the source of the signal was moving with the celestial sphere. However, the pulsation was so rapid that it did not seem possible to come from a star. In fact, it had an indisputably artificial feel to it (Fig. 4.5).

---

[1] 10^{18}.
[2] 10^{12}.

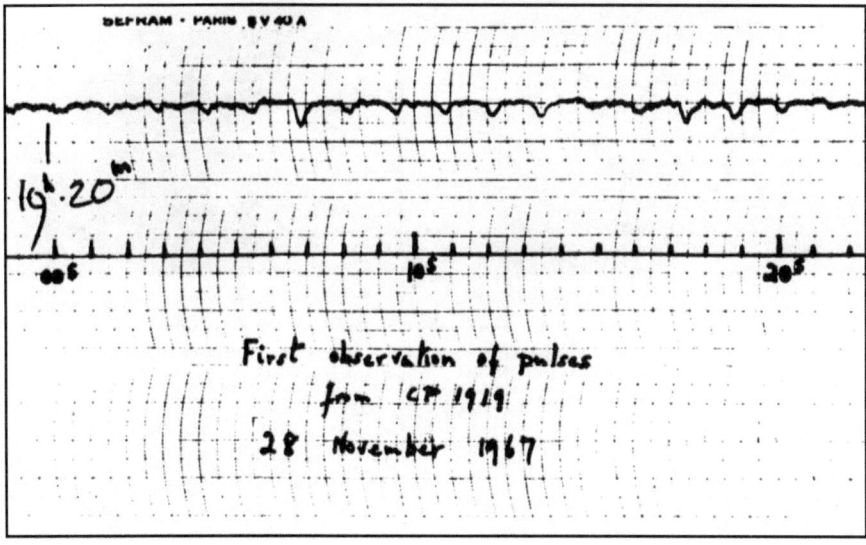

Fig. 4.5 Jocelyn Bell standing near the radio telescope of the University of Cambridge with which she found the pulsing signal of B1919+2, the first known pulsar. Courtesy of Cosmic Search magazine (http://www.bigear.org) and Annals of the New York Academy of Science

A few years before (in 1961, 1 year after the Ozma Project), Drake together with other scientists had organized in Green Bank what can be considered as the first SETI congress. In this congress it was determined that a pulsing signal would be a perfect indication of interstellar communication, easy to discriminate with respect to other signals from the galaxy. For this reason, and because no natural phenomenon was known that produced periodic pulses, an excited Bell and Hewish thought they had detected the first radio emission from an extraterrestrial intelligence, and they baptized this signal as LGM-1 (LGM: *Little Green Men*). Later they found more of these pulsing signals in the sky, and it eventually became clear that they were a natural phenomenon. The Little Green Men theory was discarded. The current nomenclature for this first pulsar is B1919+21, although it is a shame that the original fanciful designation was lost.

As metronomes in the sky, pulsars are the most stable and precise natural signals. Their discovery revealed a new specimen of stellar fauna and won Anthony Hewish the Nobel Prize for Physics in 1974. The Nobel Prize committee unfairly left out Jocelyn Bell, the student who first discovered it.

5. Where to Look

Traveling Photons

SETI scientists are convinced that, of all possible types of radiation, any signal coming to us from another galactic civilization will have the form of electromagnetic waves; therefore, the search centers exclusively on these types of radiation. Why electromagnetic waves?

Well, to start, electromagnetic radiation is extraordinarily versatile compared to other types of radiation. No other particle in nature is like the photon in relation to the ease in which it can be manipulated, detected, directed, or focused. No other particle consumes less energy when it burns and travels faster. In addition, the universe is surprisingly transparent to this radiation. Photons from remote galaxies come to us almost without any distortion. Only in the last meters of their journey does the terrestrial atmosphere interfere, masking most of the electromagnetic radiation except visible light and radio waves, for which it remains transparent. As a consequence, if we want to study other areas of the electromagnetic spectrum (X-rays or gamma, infrared or distant ultraviolet radiation), it is necessary to go out of the atmosphere and to install telescopes in orbit. Without exaggeration it is possible to say that practically all the knowledge we have of the universe has been obtained from the study and analysis of the electromagnetic radiation that comes to us from the different celestial bodies.

Only neutrinos also have this ability to cross tremendous long distances, traveling almost as fast as light. Unfortunately, neutrinos are much more complicated particles to study. To detect them, we need very large instruments that, nonetheless, only manage to detect a negligible percentage of them. To focus them, to

F.J. Ballesteros, *E.T. Talk*; Astronomers' Universe,
DOI 10.1007/978-1-4419-6089-4_5,
© Springer Science+Business Media, LLC 2010

reflect them, or to send them in the desired direction is practically impossible for us today; we still don't have a neutrino telescope. (There are detectors with this name, but they are not really telescopes in the traditional sense.)

As for other particles that might be considered candidates, like protons or electrons, provided that they have to be charged particles, the interstellar magnetic fields (due to stars or to the galactic magnetic field) have the bad habit of altering their path, so that it is very difficult to know what direction they originally come from.

The Waterhole

Even if we limit ourselves to studying electromagnetic radiation, the searching range is enormous. Which wavelengths, out of the whole range of the electromagnetic spectrum, are the most appropriate to search? Planetary scientists think that practically all atmospheres are transparent to radio waves, and a significant fraction of them are also transparent to visible light, so that both options look like a good choice. It would not make sense to use a type of radiation shielded by the atmosphere, since not all civilizations have necessarily developed craft that would fly in space.

Of both options, radio waves have the greater advantage, due to the fact that stars emit relatively few radio waves, whereas they all emit a great deal of visible light. For this reason, it is easier to detect a radio broadcast leaving a planet orbiting a star than an emission in visible light. Using radio waves, a planet can easily become more visible than its star; on the other hand, a luminous emission must be *very* powerful to stand out against the background of light from its star.

However, inside the region of radio waves there is still a wide spectrum of wavelengths to choose from. Of course, any of them could be used, and in fact SETI searches have been done using different band frequencies and widths. But there is a zone in the radio spectrum that seems to be specially favored by nature. This is the microwave window (Fig. 5.1).

The microwave window is a region of the radio spectrum where the noise due to natural causes (the noise of the galaxy, of

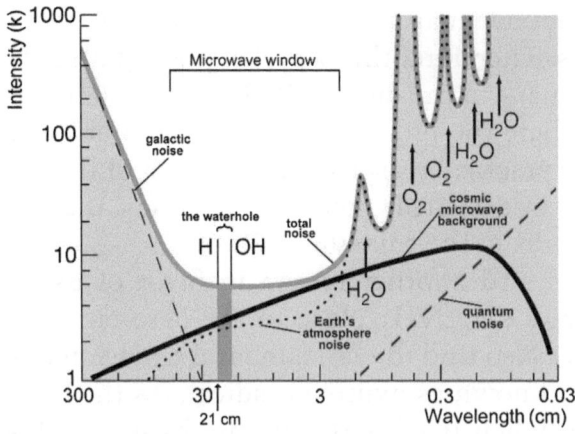

Fig. 5.1   The microwave "window" is a radio region where the contribution from different noise sources is minimal. Inside it one finds the "waterhole"

the star, of the cosmic microwave background, and the quantum detection limit) is minimal. As its name indicates, it is found in the microwave region and provides an especially silent channel that favors the reception of any artificial signal that could be emitted in this band. As we see in the graph, natural sources of noise increase enormously to the left side and to the right of this window, making it the most silent part of the spectrum for any observer in the galaxy. Even when we add the noise due to the atmosphere of Earth, we see that this region continues to be the most silent, which makes it the best frequency to be studied from Earth's surface (at least at the moment; our civilization produces more emissions in the microwave zone all the time, increasing the background noise in this zone of the spectrum).

The microwave window turns out to be interesting for interstellar communications for other reasons, too. Inside it we can find a unique standard wavelength, which must be known to any advanced civilization in the universe – the fundamental emission of the neutral form of hydrogen atoms, which have a wavelength of 21 cm. Hydrogen is the most abundant atom in the universe, and Cocconi and Morrison suggested in their article that the search should focus on wavelengths near 21 cm. The idea behind using this frequency is that it is a way to tell to the rest of the galaxy, "Hey! I'm so advanced technologically that I know that hydrogen is the most abundant atom!" This is a fact we have learned only in recent times. Therefore, this wavelength might serve as an interstellar bookmark in the dial of "Radio Galaxy." It was also in this frequency where the project Ozma worked, not for these reasons but for more practical reasons – the detector of the radio telescope they used was designed, precisely, to study the distribution of hydrogen atoms in the galaxy.

There is still another reason in favor of using the microwave window. Relatively close to the base emission of neutral hydrogen, we also find the fundamental emission of the OH molecule, better known as hydroxyl radical. As the hydroxyl radical yields water when it joins the atomic hydrogen $(H + OH = H_2O)$, the region included between both emissions has been named with the poetic name of the "waterhole," though in this zone no emission of water molecules takes place. Life on Earth is based on water, and as we have already seen there are very good reasons to think that the same thing will happen in the rest of the universe. Likewise, the same way the animals meet around a waterhole in the savannah, it is possible that these other intelligent beings based on water see also the symbolism of water = life, and consider the waterhole a meeting point appropriate among biochemical siblings.

A Question of Sensitivity

A basic component in these searches is the sensitivity of the radio telescope. The greater its sensitivity, the more capable it will be at detecting weaker sources. As we have seen earlier, the Ozma Project could only detect sources emitted by an equivalent radio telescope up to a distance of 12 light years. If the source was farther, its signal would arrive too weakly to be detected by this instrument.

The other main characteristic is angular resolution, that is to say, the smallest angular distance at which two objects must be separated in order that the telescope can distinguish them from each other. If two celestial objects are separated by a major angle, the telescope will see clearly that there are two objects, but if they are separated by a smaller angular distance, the telescope will not

Fig. 5.2   The angular resolution depends on the maximum size of the system

be capable of distinguishing both objects and will see them as a blob. Therefore, the smaller the angular resolution a telescope has, the better.

The resolution depends directly on the telescope width, or rather, on the maximum possible distance in the collecting area of the telescope. On the other hand, the sensitivity depends directly on the total surface of the collecting area (though it also depends on the detector used). In radio astronomy it is easy to have different sets of radio telescopes and to make them work in unison, combining their response to make it appear if they were a single radio telescope. On these occasions, it is common to think that the results are equivalent to that of a much bigger radio telescope. For example, two radio telescopes separated at distance L working in unison are equivalent to a radio telescope with a diameter L (Fig. 5.2).

But actually they are only equivalent in angular resolution, in their ability to separate objects that are very close. Nevertheless, the second one has a much greater collecting surface and therefore a higher sensitivity, which will make it capable of detecting sources much weaker than those that the set of two telescopes of the left side could detect.

At present, the most sensitive radio telescope of the world is that of Arecibo's Observatory, in Puerto Rico, administered by Cornell University. This radio telescope possesses the largest collecting surface in the world, a gigantic antenna that is 305 m in diameter, that is to say, a surface of approximately 73,000 m². Larger radio telescopes exist, such as the RATAN 600, in Russia, which is a ring structure 600 m in diameter, making it the largest individual radio telescope (rather than sets of multiple radio telescopes) in the world. But since it is just a ring, without any collecting surface in its interior, it has a smaller total collecting area, approximately 1,000 m², so at the moment Arecibo continues to be the largest (Fig. 5.3).

Fig. 5.3 Parabolic antenna of Arecibo's radio telescope, 305 m in diameter. The Gregorian dome is suspended from three gigantic columns, where signals collected by the principal antenna are detected. At the *right*, note the building complex at Arecibo's Observatory. Courtesy of Daniel R. Altschuler/ Arecibo Observatory

A Morse Code for Stars

Well, we have seen the advantages of radio waves for interstellar communication, and in which zones of the radio spectrum it seems most promising to search. But what do we expect to hear? What characteristics must an extraterrestrial signal have in order for us to be certain of its authenticity?

The principal characteristic that distinguishes an artificial radio signal from the signals generated by natural phenomena is its spectral width or bandwidth – how much space it occupies on the dial. Any signal with a width lower than 300 MHz will be artificial, since nature cannot generate a signal of this type. For this reason, one of SETI's principal criteria is to find narrowband signs. In addition, a narrowband has the advantage of increasing the signal-to-noise ratio: the narrower the bandwidth, the less noise (Fig. 5.4).

If the radio signal is going to be detectable across interstellar distances, it must be sent in the form of a very slow Morse code, with a bandwidth of 1 Hz or less. This means you have to work with long integration times (the times during which the detector

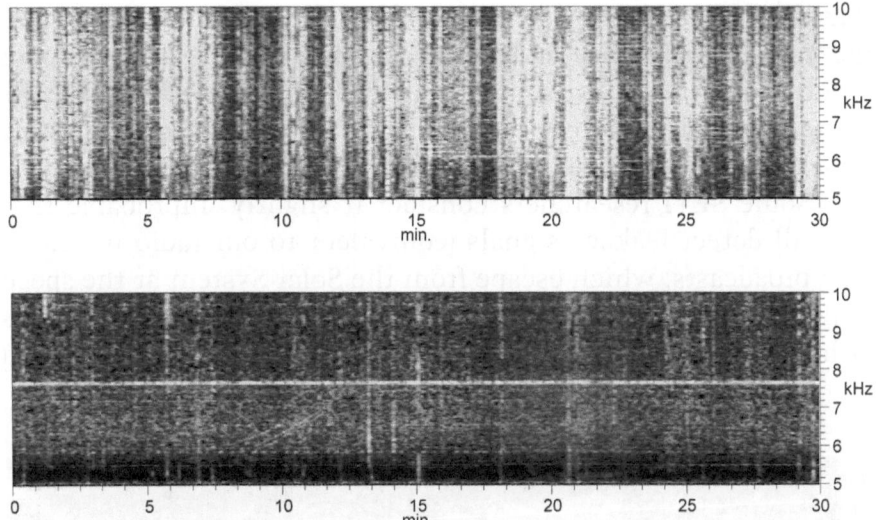

Fig. 5.4 *Top*: radio spectrum of a storm. Every lightning produces a broadband signal that spreads along almost all the bandwidth, during few seconds (time goes from left to right). Natural radio emissions are always "broadbanded." *Bottom*: an artificial narrow band signal produced by a ham radio station. A narrow band is a typical signature of an artificial emission

is measuring a signal), which improves the detection quality. If the signal were oscillating very fast, for example, 200 pulses per second, and the time in which the detector was integrating data was at 10 s intervals, 2,000 pulses of entry would all end up in one exit, with all the information they could have received being lost. In addition, if the integration time of the detector were as brief as half of a millisecond, it would be difficult to distinguish the signal from the background noise. Therefore, slow pulses are the ones to look for.

Another important condition is for the signal to be repeated. Throughout the years SETI has been operating, there have been innumerable promising signals, but on having searched for these again in the same direction of the sky, they have not been repeated (except for once, which we will later see). In many cases it has been verified later that they were coming from our own planet: interference, military radars, airplanes, telecommunication satellites, space probes, and so on. As a consequence, if these signals are designed to be detected, or if they are a leakage of the emission these civilizations are using for their own purposes, chances are that they will appear again coming from the same direction. Finding a signal only once and never hearing anything again is poor proof. In addition, if the signal transmits some type of message, its being repeated reinforces its artificial nature and avoids possible losses of information.

Some SETI researchers consider it slightly improbable that we will detect leakage signals (equivalent to our radio or television broadcasts, which escape from the Solar System at the speed of light) that are not intentional. The argument is that such types of high-powered transmissions are really energy losses, which in

Fig. 5.5 Our planet is emitting continuously radio signals that escape from our planet at the speed of light: TV programs, radio broadcasts, radar pulses... In 1906 the first radio program broadcast was emitted. Since then, Earth's radio emission has grown to the point that nowadays our planet is brighter than Sun in the radio region

time are bound to be offset. On Earth, the powerful analog trans-missions of television are being replaced with low-power digital transmissions or with optical fiber networks, so that the period in which a civilization emits these leakages is likely short. Some emission might last a longer time, for example, radar used to con-trol aircraft traffic or for the monitoring of meteorites. Nonethe-less, these researchers deem more probable the detection of an intentional signal – a message (Fig. 5.5).

A way of increasing the artificial character of an intentional signal would be to include a calling signal, something that can't occur with any natural phenomenon, for example, the principal modulation consists of a series of prime numbers: 2 pulses, 3 pulses, 5, 7, 11... (the fictitious case discussed in the introduction would also be a good example of a calling signal). A signal like this nec-essarily implies intelligence with mathematical knowledge. This would rapidly attract the attention of anyone receiving it. Over this principal modulation, inside the signal there can be more layers with codified information (sub-modulations, or perhaps changes in the polarization of the wave) where the real message might be. True, this sounds more like the plot of the movie *Contact*, which was based a book written by the well-known American astronomer Carl Sagan, who was always very involved in SETI. Sagan rigor-ously outlined in the novel what SETI researchers are still expect-ing to find in the message of an extraterrestrial civilization.

Here is a summation of the characteristics of what we expect an extraterrestrial signal designed to be detected would possess:

1. Narrowband to discriminate it from other, natural signals.
2. Slow pulses, to be easily detectable.
3. Signal repetition.
4. A calling signal with, for example, mathematical content.
5. Various layers encrypted in the signal, with the real message.

We will learn more about this in the next chapter.

6. Searching Strategies

Signals in the Sky

In SETI, the strategies used for searching for celestial signals are divided into two main groups. The first group focuses on the study of specific targets with a known location, which are, for one or more reasons, good candidates to have civilizations. This type of search usually concentrates on nearby stars, from which a hypothetical signal would arrive with more intensity than from elsewhere, and from stars similar to the Sun (since this is the only star we know in whose planetary system life has emerged).

Among these candidates, we target the single star systems, ones that have no star companion, rejecting the systems formed by two or more stars. This is because having multiple stars seems to reduce the possibility of planets being formed, since most of the material in the nebula will be consumed in the forming of the stars. In addition, any existing planet in these systems would hardly possess a stable orbit.

Priority is also given to the oldest stars, in order to allow time for complex forms of life to evolve. Hence, the most massive stars are discarded, since they have a short duration. (They explode after a few million years, in comparison with the thousands of millions of years that others last.)

Searches of this type focus on specific targets and require observing the candidate stars during long periods of time using huge high-sensitivity radio telescopes. But with this method, it is possible to study only a very limited number of candidate stars.

To complement these searches, a second type of strategy is used, consisting of carrying out indiscriminate tracking of the whole sky, looking for promising signals of unknown origin without realizing a priori considerations. Here the situation is inverted.

F.J. Ballesteros, *E.T. Talk*; Astronomers' Universe,
DOI 10.1007/978-1-4419-6089-4_6,

It is not convenient to use big telescopes, since these are only capable of observing a small fraction of the sky at a time. If we want to do a thorough tracking of the whole sky, it is necessary to use smaller radio telescopes, capable of observing simultaneously bigger portions of the sky; however, these telescopes do not have the ability to detect the weakest sources. In this type of SETI tracking, amateur radio astronomers often participate using their own small-size antennas.

A curious variant of this strategy consists of looking at wider zones of the sky, expecting to find some powerful sign. This approach was common enough in the Soviet program. It was endorsed by the theory that energy efficiency increases in a civilization as time passes, from which we deduce that most ancient civilizations must be handling extraordinary quantities of energy, which would make them easily visible. A Russian astronomer, Nicolai S. Kardashev, designed a classification of civilizations depending on this energy consumption, defining three types of civilizations. Those of Type I would be capable of handling a quantity of energy of the same order that its own planet produces (we have not reached to this phase yet). Those of Type II could use a quantity of energy of the same order a star produces. These civilizations would be capable of absorbing a good part of or the totality of their star's light to be used for their own benefit. Science fiction usually associates with this type of civilization with really spectacular astro-engineering constructions, such as a Dyson's sphere, conceived by the mathematician Freeman Dyson, which consists of a spherical cap that would completely cover the star, to avoid any loss of light to space and utilize all of the energy for itself. Or, on a more minor scale, there might be a Ring World, as the one conceived by the science-fiction author Larry Niven, a broad circular ring surrounding the star whose rotation would create a centrifugal force that could be used as a type of pseudo gravity that would make the surface habitable (Fig. 6.1).

Kardashev's Type III civilizations would have reached such an advanced level of development that they could handle quantities of energy of the same order of those produced in a whole galaxy. These civilizations would be so evident that a civilization Type III in another galaxy would be easier to discover than one of Type I in our vicinity.

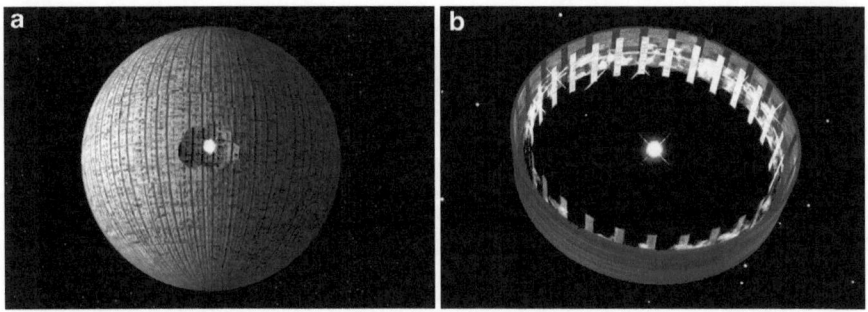

Fig. 6.1 Dyson's sphere (*left* side, showing the star in its interior) and Niven's Ring World (*right*), astro-engineering objects within reach of Type II civilizations. Courtesy of Steve Bowers

Returning to the searching strategies, one of the problems that SETI has constantly met is that it is very difficult to get enough of a radio telescope's observation time when it is aimed at a specific zone, especially when nobody is sure there is anything worth observing there. The competition among scientific projects for the right to use these instruments is very fierce, and usually observations that present more guarantees in terms of obtaining results are favored.

In addition, the ideal SETI search would be one that could combine the advantages of both strategies: to trace the whole sky, and to do it with a radio telescope of great sensitivity. Is something like this possible? Well, it seems to be so. A group from the University of California in Berkeley has managed to solve both problems with the ingenious project called SERENDIP, which was born in 1979.

SERENDIP's idea consists of putting inside a radio telescope a radio detector and leaving it there as a kind of parasite (or in piggyback mode, as the designers prefer to call it) while the radio telescope is used to do other types of observations. This way, when astronomers are studying a region of the sky in which they are interested for their own researching tasks, SERENDIP's detector will be at the same time gathering SETI information from the same region. Although the astronomers of SETI do not have any control over which area is observed, after a few years a large swath of sky will have been observed. And many places will have been observed on several occasions, which turns out to be fundamental in terms of verifying if the candidate signals repeat themselves. In 1992, SERENDIP was installed as a parasite in Arecibo's radio telescope, where it still continues gathering information of excellent quality.

Wow!

SERENDIP is one of the oldest SETI projects operating, but the record is held by the SETI program based at the State University of Ohio in Columbus. In 1973 this program began to listen to radio signals using the radio telescope Big Ear, becoming the first radio telescope to continuously look for signs from extraterrestrial civilizations. It has also been up to this moment the longest of the SETI programs, working uninterruptedly during 22 years.

In 1977 Big Ear detected the famous signal "Wow!", a particularly intense radio signal that seemed to possess all suitable characteristics – a pulse of narrowband; slow, much more powerful than the background noise; and located precisely in the surroundings of the wavelength of 21 cm. The printout shown indicates from where this signal was gathered. The time advances downwards, and every column represents a channel, i.e., a narrow interval of frequencies. The signal is circled, and next to this is the exclamation "Wow!", written by an astronomer who was analyzing the information and was struck by its intensity; the name stuck. As we can see, what we have is a narrow frequency signal that occupies only one channel of frequencies (channel 2, or the second column) (Fig. 6.2).

The curious letters that form the sign represent values of intensity relative to background noise higher than 9, since for every channel, the printed exit admitted only one character. Actually, the sequence "6EQUJ5" represents the values 6, 14, 26, 30, 19, and 5. The peak of 30 indicates that at this moment the sign was 30 times more intense than the value of the background noise. A really promising sign! However, after its detection, as soon as it was possible, Big Ear was aimed back at the same area of the sky, and the only thing it heard was silence. Since then, different radio telescopes have been pointed back at the same coordinates, but never with any results. So what *was* this signal?

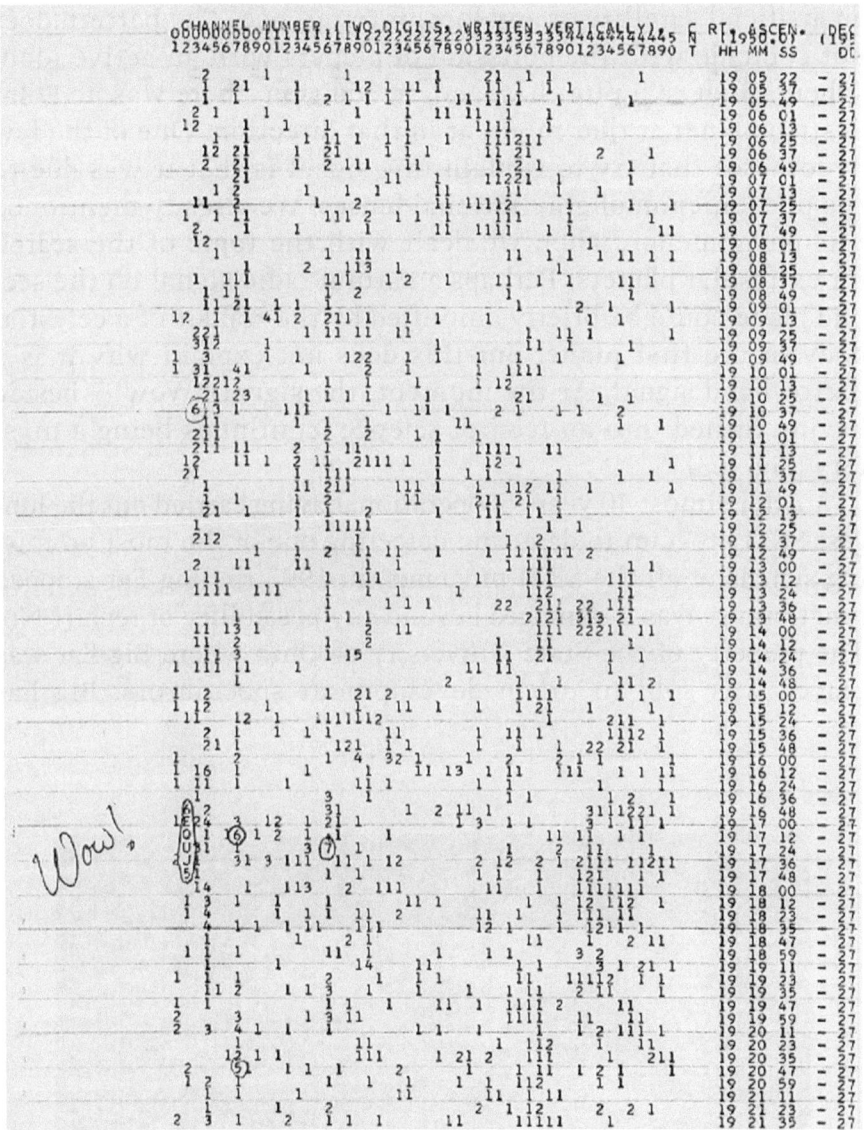

Fig. 6.2 The "Wow!" signal detected in 1977 by the Big Ear. Note that the numbers of the right column are not time hours but Right Ascension and Declination, that is, sky coordinates. Courtesy of the Radio Obervatory of the State University of Ohio/North American Astrophysical Observatory/www. bigear.org

The truth is that after all the time that has elapsed since it was heard, the signal "Wow!" has not been explained. The only thing we know for sure is that it originated at some point more distant than the Moon; therefore, it could not have been either an artificial satellite or any local interference. The pattern does not coincide with the emission of planets with an active iono-sphere, such as Jupiter has, and, in addition, there was no Solar System planet at that moment in that direction. One of the few hypotheses that try to explain this signal is that it was due to the phenomenon of gravitational lenses. We already mentioned this phenomenon when we dealt with the topic of the search for extrasolar planets. Perhaps a natural radio signal on the second plane could be briefly amplified by the transit of a celestial body in the first plane. But this does not explain why it is a narrowband signal. At the moment, the signal "Wow!", beside having turned into an icon of science, continues being a mystery (Fig. 6.3).

After almost 40 years of operation, having carried out the longest SETI program to date, and detecting one of the most promising signals of all the SETI programs, in 1997 the Big Ear stopped functioning. Was it damaged beyond any possibility of repair? No. The property of the State University of Ohio where Big Ear was placed was sold to urban development speculators. Big Ear

Fig. 6.3 Big Ear in 1977. (Courtesy of the Radio Observatory of the State University of Ohio/North American Astrophysical Observatory/www.bigear.org)

was demolished in 1998. Today, as a "monument" to its memory and perhaps to stupidity and greed, a golf course occupies its place.

NASA's SETI Program

As we have seen, the decade of the 1970s was a time of great SETI activity, with exciting moments such as the signal "Wow!". Even NASA became interested in the topic and decided to develop its own search program for extraterrestrial intelligences. A good part of the work for this was done by the SETI Institute.

The SETI Institute was founded in 1984; it was originally established to help NASA develop its program, pulling together most U.S. scientists involved in the topic. Both institutions were involved in establishing the SETI program for NASA called HRMS (High Resolution Microwave Survey), partly to offset the giggle factor that some politicians associated with SETI's initials. For several years, HRMS scientists worked to come up with a working plan, search strategies, a signal selection plan, and elimination criteria; they also worked to develop the software for all the instruments and detectors necessary for this ambitious project. Finally, the work was completed; it would be a 10-year observational program in which 1,000 Sun-like candidate stars within 100 light years were going to be studied, using for it Arecibo's radio telescope in which new microwave detectors of great spectral resolution had been installed.

On October 12, 1992, the ambitious HRMS project of NASA began its work. Arecibo's antenna was directing its attention to the star Gliese 615.1A and was observing the first candidate of the long HRMS list. But on October 1, 1993, the U.S. government canceled the funds for HRMS, and the SETI program of NASA came to a halt. Senator Richard Bryan, known for his opposition to the SETI program, gave the final thrust. In budget hearings he managed to include a last-minute amendment to end the SETI program, and the Senate voted in plenary session in favor of its cancelation. In his declarations to the press, Richard Bryan concluded: "This hopefully will be the end of Martian hunting season at the taxpayer's expense."

Why was SETI canceled? Mainly due to pressure from political groups, which accused SETI of not being science and wasting public monies. But the SETI program of NASA was indeed good science. It was an exciting scientific program that had been supported by numerous scientists around the world, including several Nobel Prize winners, who gave their full support to the program. HRMS was a rigorous project, whose scientists had obtained a great consensus on how, where, and when to look for signals. The $10 million annual investment was well worth it, bearing in mind what might be obtained in exchange if it were successful. The results derived from the program also had scientific value, since the instrumentation and methods being developed could be used in other fields of science or technology (for example, the analyzer of SETI's frequencies proved to have practical applications for air-traffic control). Still more, it was an intellectual adventure, whose educational components might draw young people towards a career in science.

Another motive for the HRMS cancelation was the expressed need for heavy budget cuts. In that year a new president, Bill Clinton, had just been elected, and the government had a budget deficit. It was necessary to cut costs in some areas. It was partly the size of the HRMS project that ironically led to its cancellation. It was small but not too small; had it been excessively small, its presence or absence would not have made a budgetary difference. Had it been an enormously big project, on which many companies were depending, it would also likely have been saved. It was just the right size, $100 million, for it to be worth eliminating and give the impression that money was being saved, especially if, as many thought, it was a useless exercise that most probably was going to give no useful results.

But it is possible that there were other motives. In some groups there is the religious belief that humankind is unique, that we are indeed the centerpiece of creation. A project whose results might question this belief becomes, at the least, inconvenient; it is better not to carry out research in the matter, at least with government funds. This possibly also explains why NASA, when preparing the rovers Spirit and Opportunity for their missions to the surface of Mars, did not include biological experiments to look for the presence of life on the planet, in spite of the pleadings of several scientists on the project.

The Revival of a Project: Phoenix

The cancelation of NASA's SETI project was a harsh blow, but it was not a mortal thrust. Actually, it served to further motivate the SETI Institute, which until then had worked as a contractor of NASA. The SETI Institute decided to continue with the project on their own and to look for private funding. A number of NASA scientists who had worked on HRMS moved to the SETI Institute, and Professor Barney Oliver, until then in charge of HRMS, began an active campaign to obtain the support of rich Californians from Silicon Valley. This work culminated in a new project that was, in fact, the HRMS project reborn, a project given, appropriately, the name of Phoenix.

One of the goals of Phoenix was to study the candidates with two radio telescopes at the same time, placed in different locations. Why? Because this way, it might easily distinguish whether the origin of an artificial signal was terrestrial or extraterrestrial. If it was only local interference being picked up by one of the radio telescopes, the other one would not detect it, resulting in the signal being automatically rejected. It had to reach both in order to be considered extraterrestrial.

Phoenix began its work in 1995, using two Australian radio telescopes: the 64 m antenna at the Parkes Observatory in New South Wales (the biggest in the southern hemisphere) and the radio telescope Mopra, a smaller radio telescope sitting 200 km to the north. In the following year, the search came back to the United States, now using the 43 m radio telescope at Green Bank (which had not been built when Drake carried out the Ozma Project), together with different radio telescopes belonging to the Woodbury Observatory. However, in both instances the antennas used together were in too close a proximity; if a signal was coming from an artificial satellite, both of them would detect it, and it would not be automatically rejected. It was better to put an ocean's worth of distance between the radio telescopes.

Finally, in 1998, 5 years after the last observation of the former HRMS project, the new project Phoenix managed to return to Arecibo, in good part thanks to the publicity brought about by the success of the movie *Contact*, which had premiered the previous year – cinema in the service of science.

But a second radio telescope was needed, and this one actually was Lovell's 76 m radio telescope, in Jodrell Banks in the United Kingdom. With the Atlantic Ocean separating both antennas, celestial signals might finally be safely distinguished from terrestrial ones. But previously it had been necessary to demonstrate this, so that daily observations started on the veteran spatial probe *Pioneer 10* (about which we will learn more in a later chapter). This probe had been launched in 1972 and was already located at an immense distance from Earth, more than 10 billion km – In other words, beyond Pluto's orbit. It was an ideal candidate to test the discrimination algorithms of the project. And the test proved to be successful. Weak signals from *Pioneer 10* proved that the technologies developed in the Phoenix project could really detect an extraterrestrial signal and distinguish it from interference coming from our own planet (Fig. 6.4).

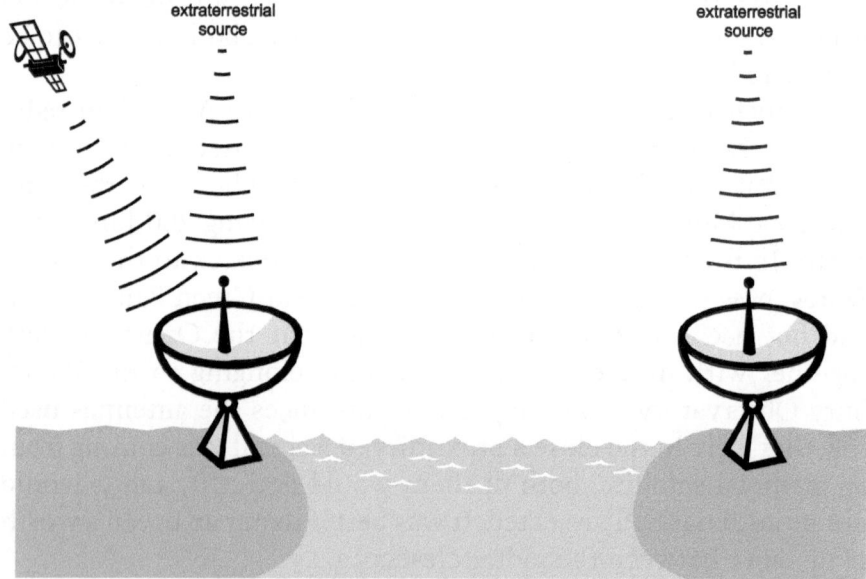

Fig. 6.4 It is difficult to know whether a signal arriving from the sky comes from a extraterrestrial source or a closer one, as an airplane. To discriminate between both possibilities, the best choice is to study the same point of the sky simultaneously with two radio telescopes. Thus, if only one detects the signal, probably it is a local source, but if both of them detect it, surely it comes from space. Nevertheless, when the two radio telescopes are too close, both could detect a local source which is distant enough, as a satellite. It is better to choose two radio telescopes that are far away from each other

Phoenix finished in 2004, having observed approximately 700 stars. It demonstrated that this methodology worked perfectly, but it unfortunately did not find any positive signal of extraterrestrial life. As for the SETI Institute, ironically, it is possible that its researchers became even more capable after they lost NASA support and resources. Not having to depend on the annual fads of the government, the SETI programs got rid of politics and consequently of the risks that the sudden loss of funds carried. The SETI has managed to be generously financed during all this time, with the support of rich patrons as Paul Allen, co-founder of Microsoft, and through merchandising. Perhaps SETI after NASA is a smaller company, but it is also more diverse, more widely accepted in academic institutions around the world, and notably more popular with the general public, as the phenomenal success of SETI@home has proved.

SETI@home

Perhaps you know of this program. Maybe you have seen it functioning on some computer or it has even been executed it on your own. It presents the aspect of a curious screen saver set to draw a few coloring graphs on the computer screen when nobody is using it. This is SETI@home, read as "SETI at home." And this is precisely what it is about, bringing SETI information to your home to analyze the data with your own computer. It is a brilliant idea to process information by means of shared calculation, behind which there is, again, the team from the University of California in Berkeley, responsible for SERENDIP (Fig. 6.5).

The SERENDIP project installed in Arecibo's antenna turned out to be so successful that it obtained much more data than it could analyze. The computers did not have enough capacity for processing the immense mountain of accumulated data, data that was still being added. In 1995 one of the team members, David Geyde, suddenly realized that there was already an immense number of computers connected to the Internet which, in principle, might be accessed. Most of these remained idle while their owners were not using them, which was most of the time. What if these computers could use this inactive time working for SETI, creating a

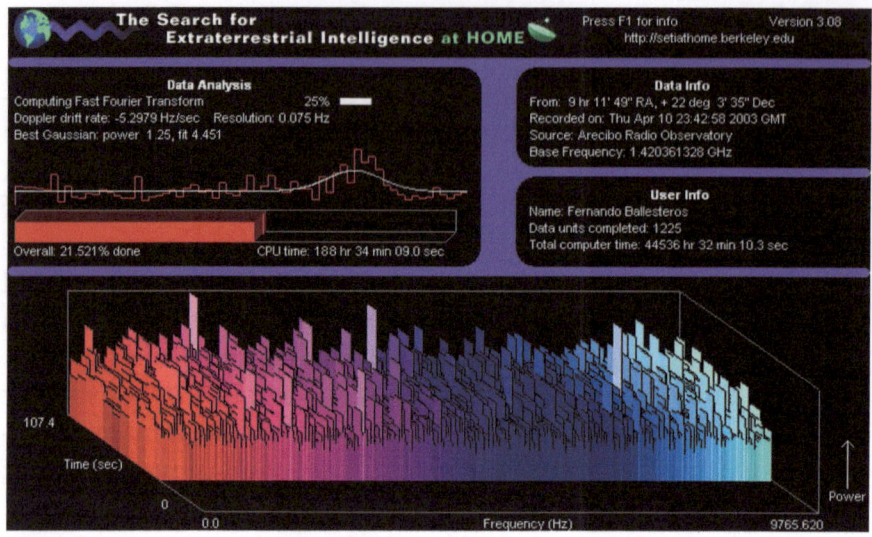

Fig. 6.5   Screen shot of the SETI@home v. 3.08 screen saver, analyzing SERENDIP data collected at Arecibo

virtual supercomputer? Then a good part of the data coming from Arecibo might be processed. It would be a solution. But how to do it?

The solution was a screen saver. Screen savers work when nobody is using the computer. This screen saver, while it worked – apart from showing nice colors – would devote itself to analyzing SERENDIP data taken at Arecibo and looking for candidate signals.[1] The program would automatically take a data packet through the Internet, and when it finished analyzing this, it would send the results of the analysis back to Berkeley, taking a new data packet and so on. So SETI@home was born, which turned into the world's first Internet shared computation project.

SETI@home started operations in May 1999. Word of mouth and people's desire to take part in the search for extraterrestrial civilizations did the rest, turning SETI@home into a success that exceeded all expectations. More than five million computers around the world took part in the analysis. It was such a big success that other similar projects followed, other shared computations also in need of calculation power – genome sequencing, protein folding, cryptography, and so on.

---

[1] There was another version of this program for Unix computers without the screen saver part, permanently working.

SETI@home worked so well that in December 2005, after 6 years of operation, there were actually more personal computers clamoring for data than data to analyze. Therefore, the original project ended, and in 2006 a new project called BOINC (Berkeley Open Infrastructure for Network Computing) took its place. As its name indicates, BOINC is no longer tied to SETI; it handles any projects that need to call on massive computation, and though it makes use of the network created in SETI@home and continues analyzing SERENDIP information, SETI is only a small part of the analyzing that BOINC performs. Nowadays it is busy processing data from projects as diverse as climatology, molecular biology, medicine, particle physics, astrophysics, and mathematics.

As for the scientific results of SETI@home, many frankly interesting signals have been found. Among them, the most promising up the time of this writing, is a signal catalogued as SHGb02+14A. By February 2003, SETI@home software had found 200 candidate signals observed more than once in the same parts of the sky and which Arecibo's radio telescope was again aimed towards, to check whether the signals were still there, or whether it had just been a coincidence (Fig. 6.6).

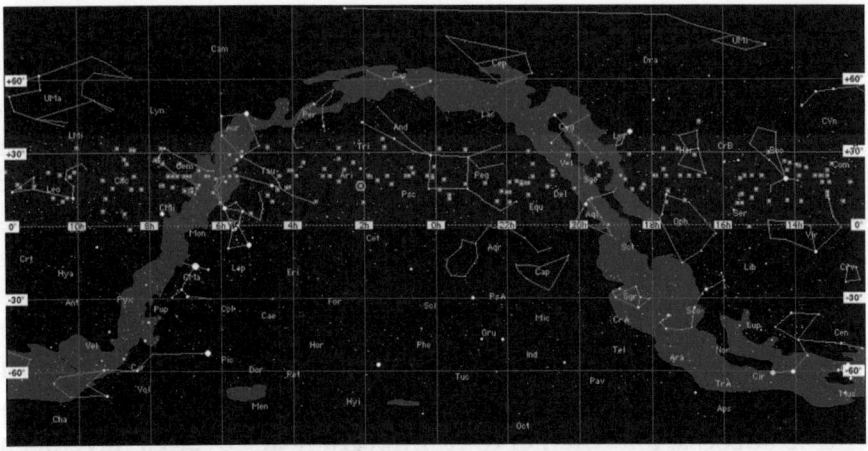

Fig. 6.6 Map of the sky showing the position of the most promising signals found by SETI@home (yellow squares with red border). In a green circle there is the signal SHGb02+14A. In dark-gray, the skyband where Arecibo's antenna can aim at. In blue, the Milky Way. Courtesy of SETI@home's project, U.C. Berkeley

When the data analysis was finished, all candidates had disappeared, except for one, SHGb02+14A. This is a narrowband signal whose wavelength is in the correct zone of the spectrum: at 21 cm, the emission of neutral hydrogen. It is located at a point between the constellations of Aries and Pisces, where there seems to be no star for at least 1,000 light years, and the signal is very weak. The radio telescope observed this signal for a total of less than 1 min, which was not sufficient to analyze it in detail, but it certainly helps to assure that it is not a matter of radio interferences or noise. Neither does it link with any astronomically known object. At the moment this signal is a real enigma, and it is not known what could have caused it.

In addition it presents another curious characteristic. Its wavelength does not remain constant but oscillates, exactly the way one would expect if the source were orbiting around something with an orbital period of only 9 days. An orbit of 9 days makes it almost impossible that it could be a planet turning around its star, but there is another possibility. Perhaps the sender could be an artificial satellite around a planet? Leaving aside wild conjectures, everything aims at SHGb02+14A as a signal as famous as "Wow!", a new icon of this search for other civilizations.

The Present and Future of SETI

Arecibo is by its own merits an icon of science, as well as an important factor in SETI's history. The most sensitive searches for signals that could come from other civilizations have been carried out from this gigantic antenna. Also from there the most famous radio message sent to possible intelligent listeners living among the stars was emitted, as we will see in a later chapter: Arecibo's message was an emotive set of ones and zeros, which in summary spoke about our biochemistry and our world. From the point of view of observational astronomy the observatori is also a great scientific success, and thanks to it, important discoveries have been achieved. Don't forget that it is the most sensitive radio telescope in the world and has the most powerful radar, and it will continue to be unsurpassed for a long time to come.

Despite all this, today it is in serious danger. In November 2006, the National Science Foundation dramatically cut Arecibo's observatory budget, which has seriously jeopardized the workings of its scientific staff as well as its ability to continue functioning. In fact, if the observatory does not manage to find funds from other sources, the National Science Foundation foresees its closing in 2011. Will we perhaps one day soon see a golf course appear in the spot where this once emblematic radio telescope stood?

OSETI

Of all the SETI projects mentioned so far, only SERENDIP in Arecibo is continuing to function as this text is written. Some new research projects have recently gotten underway, however, contributing important innovations. As we saw earlier, the universe is also incredibly transparent to visible light, and thus, this spectrum area would also be promising were it not for the brilliant glare of stars in the visible zone.

When the first SETI projects were designed, they found that for optical communication it was necessary to have an impractically strong luminous emission that would exceed the shine of the star, so that the radio search turned out to be the most practical thing. However, this changed with the arrival of commercial lasers. With this technology it is possible to generate very intense pulses of light that might even exceed the shine of a star for brief instants. This possibility opened the door for a new type of search called OSETI (Optical SETI).

Basically, OSETI is the same thing as SETI but applied to the visible range. That is to say, it deals with searching for possible pulses of laser light emitted from planets around other stars, though there are some other differences with the radio search. One of these is the limit imposed by Heisenberg's Principle of Uncertainty. We will not explain here what this principle consists of. It is sufficient to know that the quantum nature of the world imposes a lower limit for the shortest amount of time a signal can last, which is based on its bandwidth. The product of both magnitudes will be always superior or at least equal to one. For reasons of economy, artificial signals tend to approach this lower limit, whereas natural signals are always well over it.

In practice this means that if you wanted to emit a signal with a very narrow bandwidth, it would have to be a slow one, as happens with the radio signals traditionally studied by SETI (which, in addition, have the benefit of minimizing the effect of the background noise, as we saw earlier). If you wanted a very quick signal, it would have to have a considerable bandwidth. In the case of laser communications, we would find ourselves in this second situation. Besides, broadband optical pulses suffer little dispersion due to interstellar dust. For this reason, OSETI focuses basically on the search for very rapid pulses but with broad bands.

There are different OSETI projects functioning. One of them, based at the University of California, is intending to observe 2,500 of the nearby stars using a conventional optical telescope of 75 cm of diameter, which belongs to this university. This serves to illustrate OSETI's major advantage over the traditional radio-operated SETI – it does not need excessively large optical telescopes, being thus within the reach of many institutions and even some amateur scopes.

Harvard's University has also participated in OSETI. Its searches began in 1998, using SERENDIP's piggyback technique. It turns out that a telescope at this university was going to perform (for other reasons) a study of 2,500 Sun-type stars. A detector was installed as a parasite of the telescope to gather information for OSETI. After 2 years of gathering information, an average of 4 weekly signals had been detected. Nevertheless, these signals did not have any particularly artificial aspect, and it is believed that they were due to a parasitic light that had entered the sides of the detector. Therefore it was decided to switch change gears and imitate the technique of another SETI project, Phoenix in this case, observing the same thing with two telescopes simultaneously.

In this instance they had the help of Princeton University's observatory and devoted themselves to observing the same celestial sources simultaneously from both locations. In 2004 the results were published. After almost 5 years of continuous functioning, they had performed approximately 16,000 observations, and except in one case no event had been observed simultaneously by both telescopes. The exception was the star HIP 107395, but this quite possibly was a simple coincidence, because when they observed that star again, nothing was found there.

At present there are many groups and individuals who are joining the search. Various universities (such as Columbia or Case) and some observatories (such as Lick), attracted by the simplicity of the program, already include OSETI among their research work. For its part, the world of amateur astronomers is becoming increasingly interested in taking part in this new type of SETI search, the technology of which is within their reach.

New Radio Observatories

After almost 50 years of SETI research, approximately 100 searches have been performed with no conclusive results, only some stimulating moments of excitement and little data without explanation. Nevertheless, we must not be discouraged, since actually only two of our searches have had sufficient sensitivity to detect signals that could come from beyond our immediate vicinity. Of course, these are SERENDIP and Phoenix, which have been lucky enough to use Arecibo's radio telescope. Therefore, the increase of sensitivity of future radio observatories is one of the indispensable requirements for SETI's progress. To have one of these observatories dedicated totally to SETI would be also a dream come true.

This dream is on the verge of being fulfilled, since the SETI Institute, together with the State University of California, are carrying out the construction of a new radio observatory, whose priority will be the search of extraterrestrial intelligences. We are talking about the Allen Telescope Array (ATA), a radio observatory being constructed as we write in Hat Creek, California, one that could be used simultaneously for SETI *and* for first-line radio astronomical research. When finished, it will consist of 350 dishes of 6.1 m diameter, which will give it a working area equivalent to that of a single dish radio telescope of 100 m in diameter – though unlike these, ATA is capable of doing simultaneous observations of different zones of the sky.

The novelty of ATA's design is that it is made up of antennas based on commercial dishes for satellites. In view of the enormous market for these types of antennas, each 6.1 m radio telescope turns out to have a very low cost. And due to their design, new antennas can always be added, extending the capabilities of the set.

The impetus for creating this radio telescope and its name came from Paul Allen, co-founder of Microsoft with Bill Gates. Together with Nathan Myhrvold (another ex-Microsoft person), Allen donated large sums of money to the project. With their donations it was possible to develop the suitable technology to carry out the first phase of construction of the observatory, which was completed in the summer of 2006, with a starting operation of 42 antennas. In October 2007 this set began its SETI observations with a sampling of the galactic plane. However, the project still needs funds to finish its work and is open to outside donations. For example, a $100,000 donation gives your name to one of the antennas (Fig. 6.7).

There is also a low-cost interesting initiative called Argus. This project is coordinated by the SETI League. The SETI League is an educational, not-for-profit independent society, which was started by a group of volunteers discontented with the cessation of the SETI program of NASA in 1993. A good percentage of the almost 1,400 associates in the league are radio fans, amateur astronomers, amateur radio astronomers, electronics experts, and other enthusiasts (including several scientists) who want to help in the adventure of

Fig. 6.7   Allan Telescope Array (ATA). Picture by Colby Gutierrez-Kraybill.

searching for intelligence on other worlds and are currently used to help out on the different projects of the SETI Institute. At present its flagship project is Argus, a network of 5,000 stations worldwide that can carry out SETI observations. The league provides volunteers with designs of antennas and electronics and coordinates the efforts and the gathering of information. When it starts to function, it will constitute the first constant sampling of the whole sky, in all directions, in real time. At the time of this writing there are a total of 100 stations.

Finally, another giant future project for radio astronomy will be the international observatory SKA, the initials standing for Square Kilometer Array. This monster, with a sensitive area of a million square meter (14 times bigger than that of Arecibo), is expected to be functioning by 2020 if everything goes as planned, though at the moment its location has not yet been decided (the most probable candidates being Australia and, second, South Africa). Like ATA, SKA will consist of a set of antennas, but these will be arranged in a curious pattern of spirals radiating from the center, so that every spiral could work if desired as a separate radio telescope.

With all these new projects, SETI's expectations of success increase considerably. It is expected that by 2015, ATA will already have studied approximately 100,000 stars. By 2025, this number might be in the millions. By adding the capabilities of other projects such as SKA, it will be possible to study with great sensitivity an immense number of planetary systems in our galaxy. These samples might be sufficiently numerous so that, if civilizations are out there, we will probably find them in our lifetime.

Part III
How? The Language
Of Communication

7. Different Languages

If SETI produces results, and we manage to detect signals from other civilizations in the galaxy, we will have a problem. No doubt, the commotion due to this news would be enormous and its repercussions wide-ranging. But the problem we refer to here is different. Could we understand the signal? Even if this signal were designed by our galactic neighbors to be easily decipherable, could we, nonetheless, understand anything? In short, is there any common language of communication possible?

To understand the difficulty of this problem, we are going to inversely approach the topic. If we sent a radio signal to any completely alien civilization, could *they* understand it? Is there any means of formulating a message that would be completely understandable to any type of intelligence we might encounter? What problems does this attempt involve?

Human Language

According to information theory, language is a means of information exchange. Words are the key by which this information is codified in order to be exchanged. In this process, information is initially a mental representation that resides inside the brain, in our case in the form of neuronal connections and chemical signals. In order to transmit this information to another brain, we must codify it using series of rules that we call language.

Language turns a mental representation into a set of arbitrary consecutive symbols – words. These words, though already coded information, at that moment remain only in the sender's mind. It is necessary to transform them into something else, such as sounds – phonemes – so that they can reach the recipient; in short, the sender

F.J. Ballesteros, *E.T. Talk*; Astronomers' Universe,
DOI 10.1007/978-1-4419-6089-4_7,
© Springer Science+Business Media, LLC 2010

must speak. Once in the form of sound, the mental representations will travel through the air as vibrations of air molecules, until they reach the ear of the receiver of the message. There, the receiver's eardrum works at turning these sounds into electrical signals that pass to his or her brain by means of the auditory nerve; that is to say, these sonorous phonemes turn into mental phonemes.

For the process to continue, it is essential that the receiver knows the key in which the information has been codified; otherwise his or her brain will receive a confused series of sounds that will not make any sense. In other words, he or she must share the language that is spoken. If this occurs, the receiver can identify the series of sounds as words and decode them successfully, converting them into a mental representation, in the form of neuronal connections and chemical signals, which he or she will understand.

Certainly, what has been described here through the example of spoken language is the basic process of communication, and the same thing occurs with written language, with sign language, and with any other channel of communication (radio waves, paper, optical fiber,…). In all cases, in order to have communication it is indispensable that the senders and recipients share the same code.

All human languages take the word as the basic unit, a certain chain of phonemes that codify a specific piece of information for the brain, i.e., each word has a meaning. Words are the "atoms" of communication. Certainly, in different languages the same meaning is represented by different chains of phonemes. And within the same language we find variations, such as synonyms, which are different chains with identical or very similar information (student, pupil). There are also the homophones, which are similar chains with different meanings (bear, bare). Starting from the union of these words, of these bricks, and using a set of syntactic rules, we can build sentences, or whole thoughts, and perform the process of communication.

Human language would not be so powerful as a channel of information exchange were it not for another characteristic: it is a symbolic language. The ability to use symbols is acquired as our brain develops. Very young children usually refer to objects by means of words that imitate the object. For example, for a 2-year-old child, a cat is a "meow meow." Cerebral development will eventually allow him or her to use a word as a symbol of a certain object, and he or she will learn that the animal that meows is called "cat." But, even if the child uses a symbol (a word) to refer to the animal, we cannot yet speak of him or her possessing a symbolic language, since the child does not necessarily understand that the same word could apply to the whole collectivity of domestic felines. As soon as the child's cerebral development allows him or her to comprehend that "cat" refers not to an object but to a concept, we will be able to say that the child's language has evolved into a symbolic language. It has been speculated that maybe we are seeing reproduced in a few years the evolutionary process that, through hundreds of thousands of years, led our species to the acquisition of symbolic language.

The power of symbolic language is huge. It allows us to share abstract information and concepts we cannot observe in nature, to describe events or objects not present but distant in space and time. It can generate an infinite number of thoughts or ideas from a finite number of words.

With a simple example we can show the power of symbolic language. On countless occasions it has been said that an image is worth more than a thousand words, but is this true? Let's compare the words "a whale" with the photograph of a whale. You might say that, in effect, the basic concept that represents the words is also present in the photograph, and that in the latter you can observe more details than those that come to mind on hearing the words "a whale."

How can we represent with an image the words "two whales"? Logically, we would show a photograph in which two whales would appear. At the moment, it seems the images win. But we have not yet used the power of symbolic language. Let's do that now. How do we represent with an image the phrase "all the whales"? It is simply impossible. Even if we were showing a photograph with an incredible number of whales, this image could represent the concept "a multitude of whales," "9,537 whales," or "many whales," but it would never show the abstract concept of "all the whales."

Is symbolic language a mere intellectual development, a cultural product? Everything seems to point to the idea that it is something more innate in us. It is true that we need to learn words in our childhood. But it is also true that we learn them amazingly quickly. As adults we cannot learn a language at the same speed.

The ability of children to learn a language is astounding. Imagine the challenge they have to face: they have to discover the internal structure of a complex system that contains thousands of units, which in turn can be joined in an almost infinite number of combinations, only a small subset of which makes sense. In addition, the linguistic statements that they receive do not explicitly show the formal structure of sentences, which has to be deduced.

In spite of this, even very small children are able to acquire and control such an extremely complex system in a short time, at an age when they are unable to execute very much simpler intellectual tasks. And they are able to carry out this achievement even if they are not stimulated to learn it. This is Noam Chomsky's "poverty of stimulus" argument. Chomsky defends the innate existence of a kind of grammatical machine in the brain: "How is it possible that from the poverty of stimulus of our daily life we accede to a language characterized by its creativity? Is it not surprising that a child not only produces new sentences, but also understands sentences that he has never listened to before?"

Although there are some cultures, such as ours, that revere children, other cultures by contrast ignore the child until he or she can speak fluently. And nevertheless, in both cases, children acquire language at the same rate, without appreciable temporal differences and without the need of any kind of education. This poverty of stimulus argument implies that the human child has to have an innate grammatical knowledge.

A well known case where the existence of this innate grammatical knowledge was shown in a spectacular way was that of the Nicaraguan deaf-mute children. In 1977, the first public school for deaf-mute children was created in Nicaragua. Its docent methodology was mainly based on lip reading and finger spelling (where each sign represents a letter, and words are spelled out). But these deaf-mute children did not have the concept of "word," as they were never exposed to oral or written language. Thus, they were not able to understand their teachers. Nevertheless, during the recess hours in the playground, where other children played and interacted with them, they started to develop a set of signs that they used for their games and interchanges.

Soon, teachers realized that the children were communicating fluently among themselves. As time went by, this sign language gained in complexity until it became a complete and mature language, with its own rules. But the teachers could not understand their pupils, so they contacted sign language specialists at the Massachusetts Institute of Technology (MIT) for decoding this new language. What the MIT specialists found was that this language had undergone an evolution. The smallest children had used as a basis the sign language of the older children (who were the first ones to develop the language), and from this they had raised the language to a higher level of complexity, including verbal concordance and other grammatical conventions. In only a few decades, a new language had appeared, which is still spoken today.

However, there are indications pointing towards an acquired origin of speech. If linguistic capability were completely innate, any healthy human, under any circumstance, would develop some kind of language. But this is not true. Isolated individuals do not develop language. This is shown in the case of "wild children" that have grown up alone in the wild, sometimes fed by wolves or other animals. Almost all of them showed some common characteristics, such as difficulty walking, not being sociable with people, not smiling, crying, or laughing, and lacking any kind of language (vocal or gestural). Once these children had been restored to society, most of them learned only a few words (although many showed an unexpected affinity for music; for example Peter from Hannover, found in 1724, never managed to speak, but he felt intense pleasure when listening to music). Very few were able to integrate completely into society, possibly only those that had had some contact with human speech before getting lost or being abandoned in the forest. In contrast to the Nicaraguan deaf-mute children, wild children never managed to develop any kind of language. This is difficult to explain if the linguistic capability is innate. These children were mute or, at most, their "communication" consisted in simple grunts. In the forest, these children were no different than any other animal.

Furthermore, speech can also get lost (at least temporarily) if an individual is isolated for a long time. In 1704, the Scottish sailor Alexander Selkirk was abandoned on a desert island in the archipelago Juan Fernandez, after several disagreement with the captain of the galleon on which he was sailing, the *Cinque Ports*, over the poor condition of the ship (in fact, the galleon would sink afterwards). Selkirk remained completely alone on the island for 4 years and 4 months. It is believed that his life was the inspiration for the novel *Robinson Crusoe* by Daniel Defoe. In fact, the island where Selkirk was abandoned was later renamed Robinson Crusoe island. Selkirk was finally rescued in 1709 by the ship *Duke*, but he had completely lost the ability to communicate verbally. He did not remember his own language. Nevertheless, this state was transitory, and soon he recovered his language.

How do these facts fit with the hypothesis that linguistic ability is innate, intrinsic to human beings?

Of course, the answer is that human speech is both innate and acquired at the same time. Innately, we have the will to speak. Babies babble consonants and vowels, and they start to speak spontaneously, vocalizing innately without any instruction from adults (although maybe there exists some imitative component, since babies of a few months look attentively at their parents' lips when they speak). We have seen that the brain, somehow, seems to have pre-wired the grammatical rules, since even in cases with the absence of a model, a complex language can be developed (i.e., the Nicaraguan deaf-mute children). On the other hand, we must acquire the language, that is, learn the vocabulary from our environment, so the actual language that a given person speaks is acquired. Moreover, the existence of language implies its usage and development in a society, for, as we have seen, in the absence of it, such as when individuals are isolated, language does not appear to develop. Thus, the environment is also critical.

How can both factors be critical? We don't know. This is one of the main mysteries of our communication system. Sometimes this is called "the paradox of human language." We will talk again about this paradox later.

When Did We Start to Speak?

About 100,000 years ago, there was not a single group of humans on our planet, as there are today but two groups: us and the Neanderthals. We are not direct descendants of the Neanderthals (they are not our ancestors but a parallel line of evolution and had bigger brains than us). They were an exclusively European species that inhabited Europe from 230,000 years ago until 30,000 years ago, during the middle Paleolithic era. We, on the contrary, evolved at the same time in Africa. We arrived in Europe only 40,000 years ago. The present inhabitants of Europe are all, in fact, African immigrants. The Neanderthals were the real Europeans.

Only 10,000 years after our arrival, the Neanderthals disappeared. The last population lived only 24,000 years ago in Gibraltar.

The extinction of the Neanderthals is still a mystery, but it seems that these 10,000 years of coexistence of two similar species in the same place had much to do with their extinction. It also seems that, despite their great brain, intellectually Neanderthals were inferior to us. We do not know of any artistic efforts by the Neanderthals (except two questionable exceptions – a necklace found in a Neanderthal site and a supposed bone flute – but both cases are seriously disputed by experts). As far as we know, every example of prehistoric work of art we know of was created by our own species. Thus, the Neanderthal's mind would not be able to handle symbols as easily as our minds. This leads some researchers to believe that, maybe, Neanderthals did not have a symbolic language, or that they had any kind of language at all. But is it certain that they did not speak at all?

A possible answer to this question can be found by studying their ears. In our case, the frequency range we hear best goes from 2 to 4 kHz. Logically, this coincides with the frequency range that we use for speaking. If that frequency interval is the one we use to transmit information, we must be able to hear it accurately. In comparison, chimpanzees transmit almost no information by means of sounds, and what they hear best is *not* in a frequency range but in two peaks: around 1 and 8 kHz.

What about the Neanderthals? In Sima de los Huesos, at Atapuerca, a very well preserved cranium of *Homo heidelbergensis*, (a direct ancestor of the Neanderthals) was found. The cranium showed very clearly the structure of the ear bones (or ossicles). Dr. Ignacio Martinez, at the University of Alcala de Henares, built from these ossicles a model of the frequencies that *H. heidelberge*nsis would hear best. The answer was that they could hear very well in the range of 1–4 kHz. That is, this ancestor of the Neanderthals had an auditory sensitivity similar to ours. And surely so did its descendant, the Neanderthal. Therefore, if they could hear so well in this frequency range, they should essentially be able to generate sounds in that same range.

Another fossil that serves to support this theory is the hypoglossal channel, a hole in the cranium, crossed by the nerve that controls the tongue. Given that the precise control of the tongue is fundamental for speech, you will not be surprised to learn that our hypoglossal channel is much bigger than that of big apes and our distant ancestors. So what is it like in the Neanderthal fossils? Well, it is as big as ours! Surely this means that their tongues were also as adept as ours.

When we speak, our modulation of words requires our lungs to release the air very slowly, in a way that is very different from when we breathe. For this, we need very precise control of the pulmonary movement. The nerve that controls this movement passes through another hole in the bones, in this case, a thoracic vertebral channel. Again, this channel is very big in our case and small in the case of big apes and our distant ancestors. And again... in the Neanderthals it is as big as ours.

On the other hand, in 1983, a Neanderthal hyoid bone was found. This bone is the one that links the tongue to the larynx, and it was identical to ours. But what were the soft tissues surrounding it like? For this, in 2005, Dr. Bob Franciscus, at Iowa University, scanned a complete male Neanderthal skeleton, including the hyoid bone, and also had several students scanned, in order to have a mathematical model of the material surrounding the hyoid bone. Applying this model to the Neanderthal skeleton, he could obtain a model of his vocal tract. From this model it was possible to obtain several conclusions. First, the Neanderthal larynx was shorter than ours. In fact, the Neanderthal larynx was similar to the one of a modern woman. That means their voices would be more high-pitched than ours.

Second, the position of the larynx was further up than ours, and thus they would have had problems pronouncing sounds such as k, g, or ch, and the differences of the sounds among the vowels would not be as noticeable as in our case. In practice, maybe Neanderthals could pronounce only two or three distinct vowels. This has led some researchers to suggest that it was practically impossible for them to have complex communication, concluding thus that the Neanderthals did not speak. But this hypothetical smaller range of sounds does not have to imply a communication problem. Today there exist complex communication systems with only a very limited number of phonemes, as for example the Silbo Gomero (Gomeran Whistle) from Canary Islands, which has only two vowels and four consonant sounds!

This, together with the three characteristics that modern humans share with the Neanderthals (the ear, the hypoglossal channel, and the thoracic vertebral channel), indicates without any doubt that 500,000 years ago there was a common ancestor to us and the Neanderthals (maybe *Homo erectus*, for which the hypoglossal channel was three-quarters the size of that of modern humans) that already were able to perform complex vocalizations and maybe had some archaic kind of complex speech. In any case, what is sure is that our remote ancestors were already physiologically equipped with all that is needed to speak.

Maybe we can go even further back in time. In the 1990s, Derek Bickerton suggested the existence of a simpler previous communication phase that he termed protolanguage. This would not be a true language, but complex language would have evolved from it. Even more, Bickerton affirmed that this protolanguage can still be found today, as a living fossil – in the speech of children under 2 years, in the speech of wild children that have been reintegrated into society, and also in the attempts of adults when trying to communicate in a language they hardly know (for example, a Chinese and a Turk with rudimentary knowledge of English, trying to speak to each other in English). This protolanguage would be an arbitrary system of symbols but without grammatical rules, articles, prepositions, verbal tenses… consisting only of sentences of two or three words without syntax, as the following examples: Go store, what say? Want milk? You hat, etc…

The surprising thing is that big apes would share with us this protolanguage, or at least this is what several experiments seem to suggest. Some of these animals have been taught to use some communication medium, such as sign language. The sentences that these "talking apes" produce are very similar to the previous examples. If this is true, the obvious conclusion is that big apes and we humans had this primitive predecessor of language, because our common ancestor already had it. Therefore, the protolanguage must have existed at least 6 million years ago.

On the other hand, there is evidence that suggests that symbolic language is a relatively recent invention. Languages do not appear suddenly but are modifications of previously existing forms of speech. Some sounds change with time. For example, in the Latin word *bonus*, the final sound *us* evolved in Hispatia to *o*, leading to the word *bono*, and afterwards the stressed *o* changed its sound to *ue*, giving rise to the current word bueno. Other words acquire new meanings (*mouse*, besides being an animal, today is used to refer to a well know computer interface) and even lose their original meaning altogether, with only the new one remaining (*brick* meant originally piece). In this way, small local variations in language usages and phonetic trends accumulate over time, producing from a given language the slow formation of several new languages. Thus, all the Romance languages (including French, Spanish, Portuguese, and Italian) derive from Latin.

The evolution of Latin is well documented, but even if this were not the case, even if we do not know a word about the existence of Latin, we could have deduced its existence given how much alike these Romance languages are. This same deductive process can be applied to other languages. For example, the following list shows how "book" is said in several European languages: English: book; German: Buch; Dutch: boek; Danish: bog; Norwegian: bok. These words are very similar. In fact, as everyone that has studied these languages knows, the similarity is not limited to this one word. All these languages are *very* similar. The logical deduction is that, in a moment during antiquity, there was a language (in this case, an unknown language that we will call ancient Germanic), from which all these languages evolved. In fact, they are all collectively called the Germanic languages.

Let us see now how *mother* and *brother* are said in different languages (in this case, languages that are not that much alike and some of them are even extinct): Latin: mater, frater; Sanskrit: maatra, bhrataa; English: mother, brother; ancient Celtic: mathir, brathair; Gothic: modhir, brothar; Lithuanian: mote, brolis. Again, the obvious conclusion is that all these languages are somehow related. The similarities among such diverse languages show that there was a previous language from which they all evolved – the Indo-European. This was the mother language of Latin, the Germanic languages, the Celtic languages, Greek, Sanskrit, the Indian languages, Albanian,

Armenian, the Slavic languages (as Russian). In this sense, it was a very fertile language. But can we go further back in time?

The answer is yes, but as we go along the river of time, we have to be careful. Fortunately not all the words change at the same rate, but some of them are better preserved than others, mainly the words associated with important concepts, such as family or food (like milk, water…). For example, in Indo-European, *nephew* was *neput*, very similar to the current English word, or to the current Catalan word *nebot*, with the same meaning. Using the same techniques as Indo-European researchers, some linguists have followed the track of these slow-changing words and have deduced the existence of a previous language that they called "Nostratic" (meaning something like "our language"). The Indo-European would be a descendent of this previous language, and also the Uralic languages (such as Finnish or Hungarian), the Semitic languages (such as Hebrew or Arabian), ancient Egyptian, Turkish, Mongolian, and even Korean and Japanese!

Using the average language change rate, inferred from the study of the evolution of languages, linguists have estimated that this language would have been spoken 15,000 years ago. As fossil remains of this ancient language we would have words such as "bar" (meaning in Nostratic grain, seed), that can be found in the Latin word "far," wheat, in the Hebrew word "bar," grain, or the Sumerian word "bar," seed. Again let us pose the same question. Can we go further back in time?

It's true that 15,000 years is a long time, but some linguists think that it is possible to go back even further, again, following the track of slow-changing words. For example, here we have this surprising list, all of them are words that mean "water": Latín, aqua; Milcayac (South America), aka; Mapuche (South America), ako; Snohomish (U.S.), qwa; Ainu (Japan), wakka; Hawaiian, wai; Rapanui, vai; classis Arabian, wad; Russian, woda; English, water; Indonesian, vatua. And there is this other list, all of them words related to the milk, milking, or breastfeeding: English, milk; Russian, moloko (milk); Arabian, malaja (to suckle); Somali, maal (to milk); Eskimo, miluga (to suckle); Algonquin, mikolum (to suckle). These words come from all over the world!

We have now arrived at the last shell of the linguistic "onion." Those linguists defending this hypothesis think they have arrived at the original language, the language from which evolved all other languages – the mother language or Proto-World. In this language, milk was *miluka* and water *aqwad*. From the language change rate, it is estimated that this original language was spoken by a small group of humans in Africa about 50,000–100,000 years ago.

The conclusion is surprising: language originated only once, and only few thousand years ago, in the center of Africa. This is long after *Homo sapiens* diverged from other related species!

This conclusion is reinforced by the human classification studies of mitochondrial DNA, showing so many coincidences that it seems impossible both classifications (human types and languages) do not have a common basis. There are also coincidences in time. According to the studies based on mitochondrial DNA (inherited only from the mother's side), the most recent female ancestor of all human beings (called, of course, mitochondrial Eve) would have lived in Africa 100,000–200,000 years ago.

During that epoch it is estimated that in Africa there were about half a million *Homo sapiens* or even more. But the genetic studies, done using people from all over the world, prove that the current human population does not come from all those human beings but only from a small subset, a small handful of people of about 10,000 individuals living 3,000 generations ago. As a consequence of this, and despite the apparent variety of characteristics that could make you think the contrary, we, the human beings of today, are a species with very little genetic variety. Genetically we are all very much alike, almost twins. We are more alike than chimpanzees, for instance. If we take at random two chimpanzees, they have four times more genetic differences than two humans taken at random.

The coincidences between the stories told by geneticists and by language evolution researchers could make us think that the great expansion this tiny group of human beings underwent was in fact a consequence of the sudden appearance of modern symbolic language inside a small subset of *Homo sapiens*. (Their brothers, biologically also *Homo sapiens*, would instead have a more elemental, non-symbolic language.) This small step in the brain evolution of a small group would have led to a kind of linguistic Big Bang, providing this group with a great competitive advantage over other human groups – to the point that they erased from history the other competing human groups and finally conquered the whole planet.

Although the previous conclusion is certainly tempting to accept, there is an alternative that allows us to reconcile the antiquity of human speech (as it applies to the characteristics modern humans have in common with the Neanderthals, according to the fossil records) with our low genetic variability. If between 50,000 and 100,000 years ago, a catastrophic event suddenly exterminated almost all the humans beings in the world, leaving only a small group of survivors, let us say, a thousand individuals, the later human beings would necessarily all be descendents of this small group of survivors. This would explain the current genetic uniformity. And all the present languages would derive from the language this group spoke. Proto-World would not be the first human language in history, but the only one that survived the catastrophe.

In fact, there are candidates as to what this catastrophe might have been. One of the best is the Toba catastrophe. About 70,000 years ago a supervolcano exploded where Lake Toba, in Sumatra, is today, producing the largest explosive volcanic eruption in the last 2 million years. The Toba catastrophe theory, proposed in 1998 by Stanley H. Ambrose, states that the immense eruption of the supervolcano, during the glacial age, caused a volcanic winter that lowered the temperature even more, producing as a result a global massive extinction of species. Among them, all the human groups except one died out – our ancestors.

So what is the final truth? We still do not know. The origin of speech remains hidden in the mysteries of prehistory.

We have seen that symbolic language, one of the most powerful tools we have as humans, is still a bit of a mystery in several ways. It shows some peculiarities that could make us think it is indeed really exceptional, maybe an unique event. But is it really unique, or is it in fact a representative (at least in some aspects) example of what we can find in other complex communication systems?

Understanding if it is possible to have a common communication system with extraterrestrial civilizations naturally leads us to ask if human language has some universal characteristics that we will find in any complex communication system. For example, is it necessary to have a symbolic language to be intelligent? Or it is possible to be intelligent by thinking in terms of concepts or visual images? Are words the only possible mechanism of articulating an exchange of information among intelligent beings? By words we understand not necessarily a series of sounds but any kind of "atom" of information that, with certain syntactic rules, are combined to create a message. If the answer to this is yes, then we may expect that if someday we receive a message from another intelligent species, it will be made up of something that is the equivalent of words, which will be of great help in deciphering the message.

Or are we, on the contrary, a unique case? Could the use of words and symbolic language be only one of multiple ways of information exchange among intelligent beings? To answer these questions we will turn our eyes towards nature, trying to find other similar examples of communication in the animal kingdom.

Vervet Communications

If we can find animals in nature whose system of communication shares some characteristics with ours, especially if these animals show clear signs of intelligence, we will have an argument supporting the idea that we need a symbolic language for intelligent communication. This would be a question of evolutionary convergence. If this occurs it is because such common characteristics are somehow favored by natural selection and entail a certain adaptive advantage for the species that possesses them. In short, for similar problems, natural selection provides similar solutions for different species.

There are many examples of evolutionary convergence on our planet. Look at the case of wings that very different types of flying animals, such as insects, birds, extinct flying reptiles, and bats possess, or how so many different aquatic animals, including the dolphin (a mammal), the shark (a fish), or the missing reptile ichthyosaurus share the same aerodynamic form. Therefore, it is likely that if we find the same driving solutions in different animal species on Earth, we will also find wings and aerodynamic forms in the fauna of other worlds. Likewise, if in our world we can find common characteristics in communication systems among beings with a certain intelligence, it will be more likely that we find the same characteristics in the communication system of an intelligent extraterrestrial species.

As we have seen before, one of the defining characteristics of human symbolic language is the use of words. Do we find something equivalent to words in animal languages? The vervet monkey is a small cercopithecus with a black face that lives in the African savanna. Vervets form a group of primates that have a very cooperative social structure. As with other species living in open areas, when vervet monkeys are eating in the savanna, there is always one on watch, up on a high place, in order to locate possible predators and to alert the others in time (Fig. 7.1).

When the lookout monkey locates a predator, it gives a shout, warning of the danger, and the others run to shelter. But unlike other animals, which cry a generic shout meaning "danger," vervet monkeys produces three different shouts, depending on which predator approaches: "leopard," "eagle," and "snake." If a leopard approaches, the watching monkey emits the corresponding shout, and all the rest run to trees and jump up on them to avoid the predator. If the emitted shout corresponds to the snake, the vervet's behavior differs: they stand on two legs and observe the grass, trying to locate the snake. And if the shout is the one corresponding to the eagle, they go fast down the trees to find shelter among the roots.

Fig. 7.1   Vervet monkey: a notable chatterbox. Courtesy of William H. Calvin, University of Washington

The example of vervet monkeys is classic in the studies on evolution of human language, since it reveals a group of beings that use a sort of primitive code, words whose meaning they clearly seem to understand. Notwithstanding, some researchers believe that we are actually seeing a simple reaction-sound behavior. Maybe vervet monkeys automatically jump up trees when they hear the shout we have labeled as "leopard," with no real comprehension that a leopard is approaching. But even if this is true, it is quite likely that human language emerged exactly this way.

Readers can think that perhaps the example of vervet monkeys is a biased one, and this curious similarity to human language is not so amazing, since these monkeys are actually close relatives of humans. Probably what we have seen is not a real example of linguistic evolutionary convergence but simple heredity shared from a common forefather. If this is so, the similarities will be greater the closer an animal is to human. But is this happening?

Our Cousins the Apes

The answer is no. We have not found anything similar among the big apes. This is surprising, since we are dealing with our nearest relatives, and they show evident signs of intelligence. Gorillas and orangutans successfully pass the mirror test, a simple test that consists of putting an animal opposite to a mirror where its image is reflected. That the animal recognizes itself in front of the mirror is considered proof that it is conscious of itself, a characteristic that we usually associate with the possession of intelligence.

Most animals fail in this test; they are not capable of recognizing themselves in the reflected image, which they usually ignore, or treat as if it were another animal. The same thing happens with very young children: they act as if they see another child ahead. We are only beginning to recognize ourselves in the reflected image at about two, curiously coinciding with the acquisition of symbolic language.

How do we ascertain that big apes indeed recognize themselves in the mirrors? A habitual trick of researchers consists of painting spots on their bodies, in zones the ape cannot directly see, and observe if they use the mirror to locate them. Big apes not only do this successfully, but they also use mirrors to extract food

Fig. 7.2 Chimpanzee studying its reflection in a mirror at the Cognitive Evolution Group premises, University of Louisiana. The big apes are among the few animals that recognize its reflection in a mirror. This so-called mirror test is considered a measure of self-awareness. Courtesy of the University of Louisiana

bits from between their teeth, and even to tidy up. In other words, they recognize themselves in the reflected image, and there is a general agreement that, as humans do, they possess self-awareness (Fig. 7.2).

Nevertheless, they do not use words. Furthermore, their vocal expressions are rather laconic and coarse. Therefore, scientists have attempted to investigate if the minds of these primates already have humanlike-precursor linguistic and intellectual structures, and thus they can be taught non-vocal languages, the most popular being sign languages for the deaf and mute. In this respect, diverse experiments have shown surprising results (Fig. 7.3).

The most famous "speaking" ape may be the female gorilla Koko. In the 1960s, scientists from Stanford University started to teach American Sign Language (ASL) to this young gorilla, born at a San Francisco zoo. Today Koko knows more than a thousand signs that she fluently uses. Scientists working with her maintain that she understands these signs, and her actions are consistent with their use. Furthermore, she has proved her ability to understand

Fig. 7.3 Big apes such as the chimpanzee or the gorilla, trained in the language of signs, seem to be capable of developing unexpected linguistic aptitudes. Courtesy of William H. Calvin, University of Washington

abstract concepts such as the future. She is one of the few animals documented to have kept pets, having taken care of several cats.

Also remarkable is the case of Washoe, a unusually intelligent female chimpanzee (females have normally shown more skills than males in these experiments), who was taught American Sign Language by researchers at the University of Nevada. Washoe learned about 800 signs, and she seemed to be able to invent new words by combining known ones. For example, Washoe defined duck as "water bird." Moreover, she even taught sign language to her own offspring. Apparently, other chimpanzees that were together with Washoe who also learned ASL communicated in it and spontaneously taught sign language to others.

These experiments have shown evidence that within the minds of these apes there exist primitive semantic structures similar to ours (maybe Bickerton's protolanguage) that reveal a "mind theory," that is to say, these apes are capable of understanding and reflecting upon others and themselves. Also, they show ability to use the equivalent of words, since they associate each sign with a meaning.

Nonetheless, it is extremely surprising that animals with these aptitudes do not use them in their natural environment because, if they have the capacity to *learn* a language, they should also have the capacity to *develop* a language. It is extremely interesting to note that these speaking apes lack a feature that seems essential if we are to accept that they fully understand sign language. When they receive new information they passively accept it and do not ask about it. Moreover, they rarely use sign language spontaneously; they use it mainly when requested by their trainer.

Thus, these experiments are not free from controversy, and some scientists believe that results are partly influenced by the researchers' interpretations. After examining the recordings of some of these experiments, it has been proved that some ambiguous signs were recorded as valid, and that researchers explained confused results as "metaphors." In other cases, a researcher's behavior unconsciously drove the ape towards the expected result. For example, the chimpanzee Lana defined orange as "an orange-colored apple," a captivating example that clearly shows the linguistic abilities of chimpanzees. But the dialog between chimpanzee and researcher (Tim) was quite suspect, as the researcher was the one who first talked about color:

Tim: What color of this?
Lana: Color of this orange.
Tim: Yes.
Lana: Tim give cup which-is red.
Tim: Yes.
Lana: Tim give which-is shut? Shelley give?
Tim: No Shelley.
Lana: Eye. Tim give which-is orange?
Tim: What which-is orange.
Lana: Tim give apple which-is green?
Tim: No apple which-is green.
Lana: Tim give apple which-is orange?
Tim: Yes.

The Amazing Cephalopods

We share so many characteristics with apes that a more or less fluent communication with our cousins seems feasible. In any case, we are such close relatives that these studies actually tell us more about ourselves and how we became humans than about the universal characteristics of communication among the intelligent beings we are searching for. For this reason, we are going to make a qualitative jump, and focus our interest on a family of surprising invertebrate animals: the cephalopods. It is probably worth a warning to readers about the gastronomical consequences of reading this chapter, because it is likely that after reading it those who like to eat squid will see them with new eyes.

It has been said that cephalopods are the creatures closest to extraterrestrial beings on Earth. These mollusks have developed their intelligence in completely different ways from vertebrates. We belong to different phyla, and our last common ancestor was probably some kind of soft blind worm that lived some 600 million years ago. But, in spite of being close relatives of snails, they have so many points in common with us that they have been awarded the title of "honorary vertebrates."

Very similar to those of vertebrates, in spite of their completely different evolutionary origin, cephalopod eyes are a perfect example of evolutionary convergence. The same thing happens with their complex neural system, which has evolved in parallel and shows a well-developed brain. This complex brain provides them with an extraordinary memory and learning ability, allowing them to easily solve mazes and difficult problems and to learn from their peers. Their tentacles provide them with an extraordinary manipulative ability, resulting in a powerful combination of abilities, and thanks to these they can successfully face challenges we imagine only superior vertebrates could face. For example, all it took for an octopus at the Munich Zoo to figure out how to unscrew the cover on a flask with food inside and pull out the food was to watch how its trainers opened the flask.

Their brain complexity also results in a very sophisticated communication system. It is very well known that cephalopods can change the color and tone of their skin (in some species even the texture) to mimic their environment (Fig. 7.4). This characteristic, which no doubt was very helpful for their survival, is also the basis of their communication system, which uses color coding combined with body postures. A certain pose, a pattern of colors and spots, or both, have a definite meaning. Thus, for example, cuttlefish turn literally black with anger. Some researchers think that posture may

Fig. 7.4   The same cuttlefish in three different states. Cephalopods show a great ability to change the color, pattern, shape and texture of their bodies

be useful to convey the basic message transmitted by the spot pattern. The number of different communicative elements is large, maybe close to 100 (depending on the species), and it is undoubt-edly much larger than that of apes, so they must have a lot to talk about. Cephalopods have intense "conversations" by combining the communicative elements into sentences, although it is not clear whether there is any kind of syntactic rule behind these (Fig. 7.5).

Cephalopods fail when confronted by the mirror test. They do not recognize themselves in the reflected image. Even worse, they do not understand the reflected image! When shown the reflected image of something edible, they ignore it, contrary to what they do when food is directly presented. This paradoxical behavior is due to the fact that their eyes are able to detect the polarized state of light (whether waves reaching the eyes vibrate

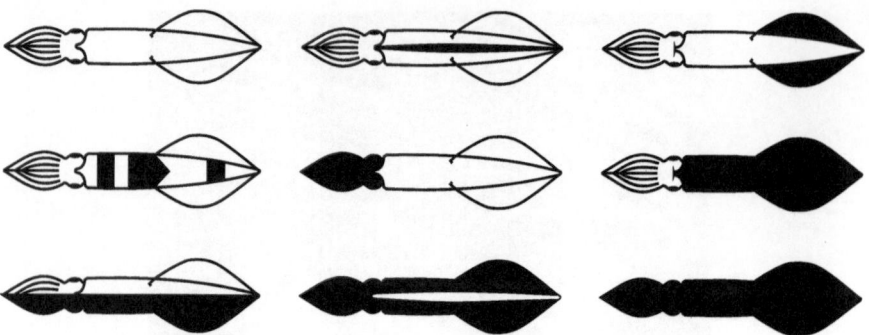

Fig. 7.5   Different examples of body spot patterns used by squids when communicating. Drawing made from Cornell's Marine Studies Center data

horizontally, vertically, or in another way). Indeed, light happens to change its polarization state when reflected on a mirror. Therefore, although direct and reflected images look the same to our eyes (which are *not* able to detect the polarized state of light), to cephalopods it is clearly quite different. For this reason, the mirror test is not conclusive in the case of these animals, because to them, their image reflection in the mirror can look very different to their peers, and it can be impossible for them to recognize themselves.

The Dolphins

In our study of communication systems between intelligent beings, one must inevitably refer to a family of vertebrates that shows evident signs of a developed intelligence, the Delphinidae. Dolphins are regarded as the most intelligent animals. Any indicator designed by humans to estimate human intelligence, and to "demonstrate" that we are the most intelligent beings on Earth, shows that dolphins can score as well as us, and can score even better in some cases (Fig. 7.6).

Just to start, the dolphin's brain is larger than the human one. This should not in itself be an indicator of intelligence; the brains of some other animals (such as elephants and whales) are much larger than human brains, even four or six times larger than ours. But we should take into account that these brains are this

Fig. 7.6   A bottle-nose of the U.S. Navy, leaps out of the water with a localizer device attached to its pectoral fin. These dolphins are trained to conduct deep/shallow water mine countermeasure operations to clear shipping lanes. Courtesy of the U.S. Navy

large because they must control a very large body. Thus, the ratio between brain and body sizes is usually regarded as a better indicator of potential intelligence. With this in mind, we beat most of our competitors and rank quite high in the intelligence scale. Nevertheless, dolphins are very close to us, as their brain/body size ratio is almost the same as ours.

Another indicator we can use is brain energy consumption, which gives an idea of how intensely the brain is working. This consumption is easy to measure by the amount of blood flowing to the brain, or by the heat it gives off as a result of its activity, which is very high in humans (one-fifth of the energy consumption of our body is due to brain activity). But, if we compare our scores to dolphins, again, we are in a tie. Apparently, their brains are as busy as ours.

There is yet another physiological characteristic also frequently employed to demonstrate how intelligent we are, and this

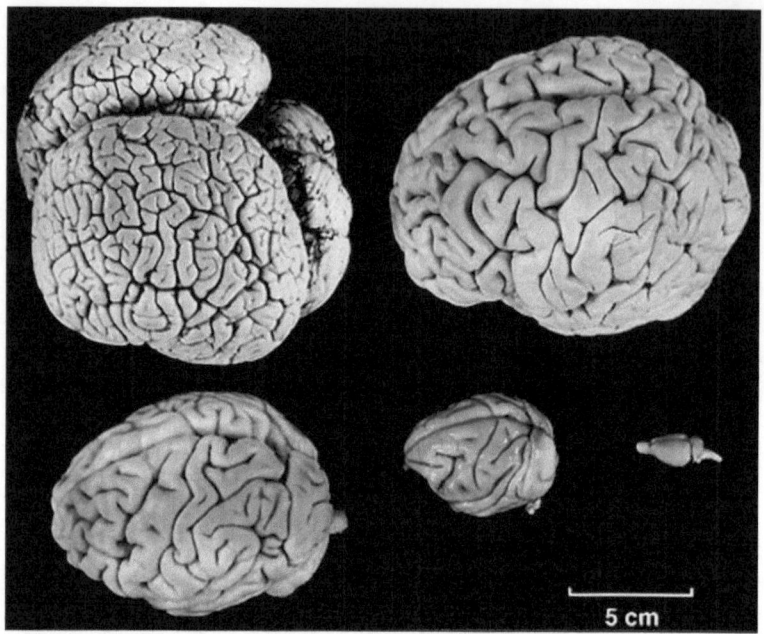

Fig. 7.7 Comparison of brains from different mammals. On *top*, the bottlenose dolphin (*left*) and a human being (*right*); at the *bottom*, from *left* to *right*, brains from a chimpanzee, a macaque, and a rat. Courtesy of brainmuseum.org and the U.S. National Science Foundation

is the number of brain circumvolutions. It is known that highest cognitive functions, such as conscious reasoning, speech, or sensory processing in the case of mammals, take place in the neocortex, that is, the external surface of the brain. The human brain has a large number of creases and folds on this external surface, called circumvolutions, which enlarge the neocortex surface, increasing the processing capacity of those functions. Well, in this case, human beings are second, as dolphins win by far. Their brains have far more circumvolutions than ours do, and it is shocking just to look at a dolphin's brain and recognize its complexity (Fig. 7.7).

Therefore, we can conclude that dolphins are actually *very* intelligent beings. But there is no real need to use the previous physiological indicators to understand that. The study of their behavior in nature and laboratory tests prove it. They can organize themselves to perform group activities, i.e., killer whales (in fact these are Delphinidae) coordinate their seal hunts through the exchange of sound messages. Different lab tests have shown that they are able to solve very complex logical problems. During the

1990s, several researchers also demonstrated that dolphins pass the mirror test, that is to say, they recognize their reflection and show clear evidence of having self-consciousness.

It is instructive to recount here a classical experiment that shows the intellectual abilities of these animals. This experiment is of particular interest because the researcher tried to demonstrate exactly the opposite of what he really showed. The experiment is known as the 1964 Jarvis Bastian experiment. The aim of the test was to prove that dolphins were not able to transmit abstract information. The subjects of the experiment were a male and a female bottlenose dolphin called, respectively, Buzz and Doris. The dolphins were put in a tank that contained two levers at one end, which were connected to a food dispenser, and a light at the other end. Depending on whether the light flashed or was on steadily, they had to press one lever or the other to obtain food. If the wrong lever was pressed, nothing would happen. After a training period, both dolphins learned to press the right lever according to the light.

After successfully passing this learning period, the second phase of the experiment began. The tank was divided into two parts by an opaque wall, which did not allow one to see through it; only sound could pass. Buzz was put in one half of the tank, Doris in the other. Thus, Doris could move the levers, but could not see the light. Because the wall allowed the sound to travel through it, if Buzz could tell Doris whether the light twinkled or otherwise, or alternatively what lever to press, the number of successful attempts on getting the food should be high, whereas it would be low if the dolphins were not able to transmit complex information. To Bastian's surprise, 96% of the attempts were successful!

The third phase of the experiment consisted of replacing the wall by a soundproof one, so that from one side to the other of the tank neither image nor sound could travel. Had there been any kind of exchange of information among the dolphins, the success rate at this point should drop to 50%, equivalent to randomly pressing the levers. On the other hand, if the number remained the same, it would mean that Doris had a different way of getting information other than the sounds produced by Buzz (unless maybe she could see the light reflections in an area not identified by researchers). The result was that, after using the soundproof screen, the success

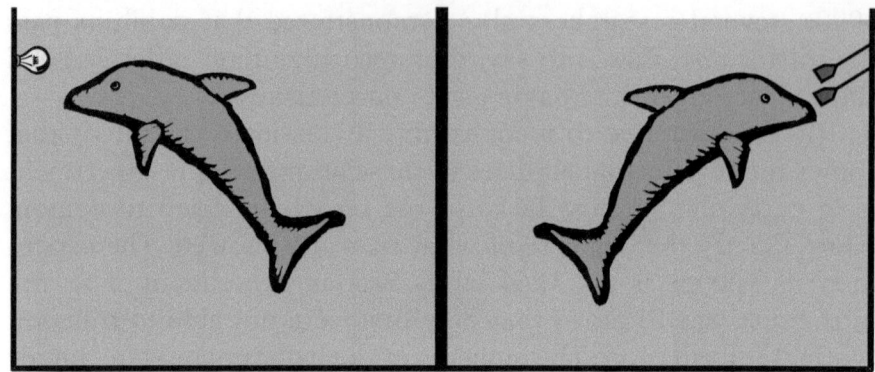

Fig. 7.8   Scheme of the 1964 Bastian's experiment. The central wall is opaque, and it can be soundproof or not

rate dropped to 50%. So, was there any transmission of abstract information between both dolphins? (Fig. 7.8)

The system of communication dolphins use is very complex. This is why numerous researchers study it. One of the most interesting studies on this system of communication was the one carried out by the Russian dolphin expert Vladimir Markov in 1990 (not to be confused with the mathematician Andrei Markov), who analyzed the vocalizations of bottlenose dolphins using tools from information theory. His work showed that in dolphin communication some chains of sounds appear grouped and form stable blocks. These blocks also present well-defined limits, marked with pauses, and are used as independent entities, combining in major structures. That is to say, these blocks behave like words in human language: blocks of stable sounds combined to form sentences with which we communicate. Therefore, it seems tremendously tempting to identify with "words" these blocks with which dolphins communicate.

There are even more similarities. The younger the dolphin, the simpler the structures formed by combinations of "words," and also the smaller the number of "words" the dolphin uses, as happens with human beings. The language of dolphins becomes more complex when it matures. And, if in effect, as Bastian's experiment suggests, dolphins are capable of communicating abstract information such as "ignition" or "blinking" or "left" or "right," we can infer that they possess some sort of symbolic language. But not only this. They also share with us something we deem exclusively human. We know that most dolphins, including bottlenose dolphins, use proper names – the only case we know of besides humans. Baptism of dolphins begins right after their birth. For several days after childbirth the mother dolphin begins to call her baby with a typical hiss (perhaps so that the baby dolphin will learn it), which will identify it all through its life, and other dolphins will use that sound to call it – its name.

8. Different World Views

When we study communication systems in other animal species, we are led to the conclusion that extraterrestrial intelligent beings are likely to have systems of communication similar to the human one in at least some aspects. That is to say, communication is accomplished by means of the combination of certain information elements playing a role analogous to words, or that the system is a symbolic language (or even both things simultaneously). But in communication, it is also very important how information is transmitted (pictorially, by means of ideograms, letters, etc.), which is going to be strongly determined by the species' specific perception of the world.

For example, would any of the previously discussed animal species be capable of understanding a two-dimensional image, a simple drawing representing a person? Probably the big apes would, since it seems that their perception of the world is similar to ours. But cephalopods possess a communication system so different that, even though they are really an intelligent species, human communication with them would probably be quite difficult. Our world views are very different. The mere fact that their eyes are sensitive to polarization results in very different understandings of the world.

Even dolphins, despite being mammals, mainly see the world through sounds. In addition to ordinary sounds used to communicate, members of the dolphin family have developed a very sophisticated system of echolocation that enhances hunting in turbid waters or without light. They emit a series of clicks from an organ called the melon located in the forehead, and get the sound through their jaws, which leads up to their ears. This system has such a high precision that it can detect a 1 mm-thick cable. This natural sonar endows them with a sort of three-dimensional X-ray vision, and they are capable of perceiving fish behind certain obstacles. Without

F.J. Ballesteros, *E.T. Talk*; Astronomers' Universe,
DOI 10.1007/978-1-4419-6089-4_8,
© Springer Science+Business Media, LLC 2010

eyes, a dolphin would not be capable of distinguishing a drawing on the surface of a closed box. Its sonar would only detect the smooth faces of the box, though the animal would indeed be capable of discerning if the box contained something or was empty.

Therefore, the characteristic world perception human beings have is bound to cause a practical limitation if we try to communicate with other intelligences. We need to be aware of this when designing a really effective system of communication. For example, we all are capable of recognizing a photograph of our environment as soon as we see it – objects in the background or in the distance (smaller), the woman in the foreground, the clothes she is wearing, which are evidently different from the physical body carrying them, and so on.

Therefore, it may seem obvious that if we want to make an extraterrestrial being understand what a human being is like but we cannot physically reach him or her, the most effective thing is to send a photograph. Nevertheless, as we have seen with other animals, success is uncertain. The photograph of a person is not a person: it is a two-dimensional surface, whereas the real human being occupies a three-dimensional volume; it is composed of spots with different colors, or even gray tones in the case of black and white photographs; and in general it has a size different from that of the human being it represents. As human beings, we manage to understand the image of a photograph thanks to how our eyes and brain work.

A World of Colors

The great majority of photons in the electromagnetic spectrum are invisible to us. Only when they have between 0.25 and $0.5 \cdot 10^{-18}$ J are they the correct energy to activate the rods and cones of our retina (light-sensitive cells), and we can see! Specifically the latter, the cones, are responsible for the sensation of color. Photons in the red zone of the visible spectrum, with less energy, activate a special type of cone called L; those a bit more energetic (in the green zone of the spectrum) activate other cones called type M; and the most energetic (blue zone) affect those of type S. Through the optic nerve these three cones send messages to the brain, which interprets them as the colors red, green, and blue.

The beautiful colors of the world surrounding us are, therefore, a "re-creation" our brain builds out of the information from the cells of the retina. They are not real; they do not exist. It is the brain that produces them, and for this it uses a curious mathematical algebra baptized with the exciting name of color algebra. The proof of this is how we see the rainbow. As we increase the frequency of the vibration of light, the different colors of the visible spectrum appear: red, orange, yellow, green, blue, and purple (we will describe it this way; it seems only Newton's eye was seemingly capable of distinguishing between indigo and violet). In the spectrum, orange is between red and yellow, and in effect when we mix red with yellow we obtain orange. Green is between yellow and blue, and the combination of yellow and blue yields green. But purple, which ensues from the combination of blue and red, is at one end of the spectrum, and is not between blue and red. It would only appear to be so if we folded this list of colors in on itself as in a ring, as it really happens in our brain (as color algebra also shows) (Figs. 8.1 and 8.2).

Another proof that colors are mental elaborations is the existence of complementary colors. Why does there exist in nature two colors that are complementary? Why is green light, which corresponds to light vibrating with a frequency around $5.6 \cdot 10^{14}$ Hz, necessarily complementary to red light, corresponding to a vibration around $4.6 \cdot 10^{14}$ Hz, and not of some other different frequency, as for example $12 \cdot 10^{14}$ Hz, in the ultraviolet zone?

The answer is that green light is not complementary to red light, but it is the mental representation of green color that is complementary to the mental representation of red color, always according to color algebra. Therefore, when a surface emits only light with a frequency corresponding to green color, we will evidently see it as green. But if this surface emits light with all visible frequencies *except* the ones corresponding to red, we will as well

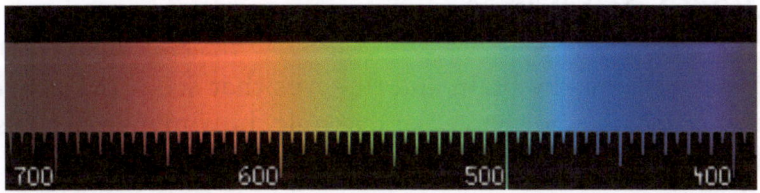

Fig. 8.1 Spectrum of visible light

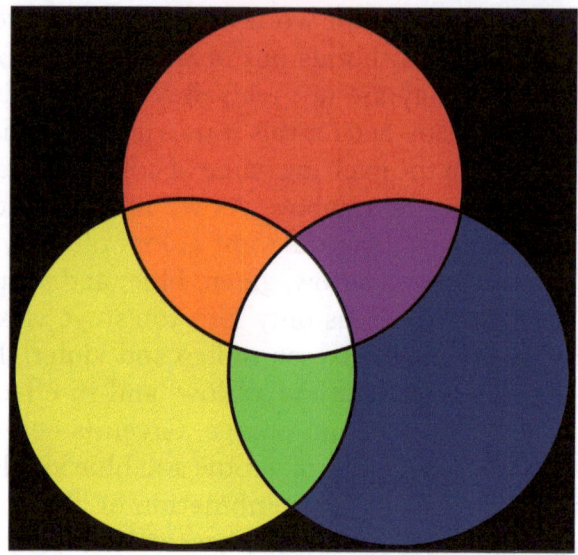

Fig. 8.2   Color algebra

see it as green, though it is actually a spectral emission completely different from that of the previous case, precisely because in our brain red and green are complementary colors. Furthermore, if the surface emits light in only two pure colors, for example in the frequencies 5.2·1,014 Hz and 6.4·1,014 Hz, corresponding respectively to yellow and blue, it will also appear to us as green since in color algebra yellow + blue = green.

   In the context of communication with an extraterrestrial intelligence, it is necessary to take into consideration how we see colors. A color photograph that we see with the same colors as the original can be completely incomprehensible to a being with another system of color receivers or with another mental color algebra, for that being may see it with colors which, for that individual, do not correspond to those of the original.

A World of Shapes

Color depends on the brain and the eyes that see it but also does the estimation of the shape. For example, how many triangles are there in the following image? (Fig. 8.3).

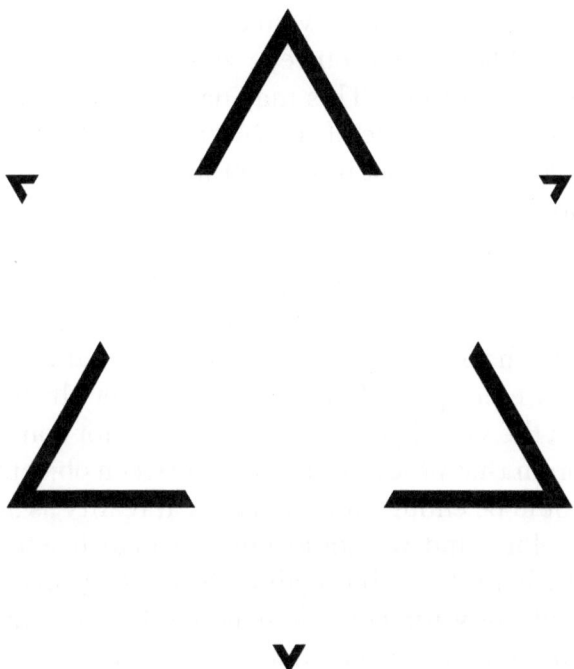

Fig. 8.3 Kanizsa's triangle

The most common answer is two, followed by six. The truth is that there are none. Nevertheless our eye insists on completing the lines and joining with straight lines the six vertexes. If we look attentively at the image, we will soon perceive a few emerging lines that are not actually there. Indeed, the reality we see is not a direct representation of what comes to our eyes because our brain (specifically our visual cortex) processes the observed images before they come to our awareness. In the visual cortex we have neurons specializing in the detection of special patterns. A group of these neurons "ignite" and trigger a neuronal response when our eyes observe straight lines. In addition, they are specialized by inclination; some of them respond only to horizontal ones, others only to vertical ones, and others only to diagonal ones, etc.

When alignments of objects appear before our eyes (as in the previous drawing), line-detecting neurons also become activated. Even when there is no actual line, our visual cortex responds as if there were. Therefore our brain tends to complete the pattern, and we guess two triangles.

This ability of the human visual system to detect alignments can be observed best when our eye works to the limit of resolution or in poor conditions. This may have negative consequences, as in the sadly famous case of the Martian canals, where Percival Lowell's eye aligned almost imperceptible elements, which indeed were unrelated, acquiring in his mind the appearance of big works of hydraulic engineering.

Our visual system also recognizes edges and contours. We see a whole composition made up of separate characteristics: borders, shapes, colors, shades, depth, and so on. Each characteristic is processed by a different part of our brain, and later all these are integrated during the visual perception process. If for some reason the part of the brain that processes borders between objects were damaged, this would be enough for us to look at reality as a continuum of spots of colors, and we could not distinguish where a person finishes and where the wall on which he or she leans begins, even if our eyes and everything else were perfectly working.

We even possess neurons that specialize in recognizing faces. In a study performed with macaques, electrodes were placed in their inferotemporal cortex. Scientists found individual neurons that showed an intense response when images depicting the face of a monkey were presented to them. The response was slightly lower when scientists presented a human face, still lower when the drawing of a smiling face was shown, and there was no reaction at all when a random pattern of lines was presented to them (Fig. 8.4).

Due to this, we can easily see faces even when they are not really there, as in the following images, where we can see microscopic pictures of the endoplasmic reticulum of a mouse's optical nerve, an onion grain, and a picture from the *Viking* orbiter showing the "face" on Mars (Fig. 8.5).

All human beings share the same brain structure. This is why all we recognize faces in these objects, although it is obvious that they are not faces, nor have any relation whatsoever to faces. They are tricks produced by the visual perception mechanism of our brains. Why do we have those cerebral structures for face recognition? Because recognizing faces has a clear selective advantage: it is better to see a face even when it is not there, than to be unable to recognize a face when it is there. In other words, it helps to

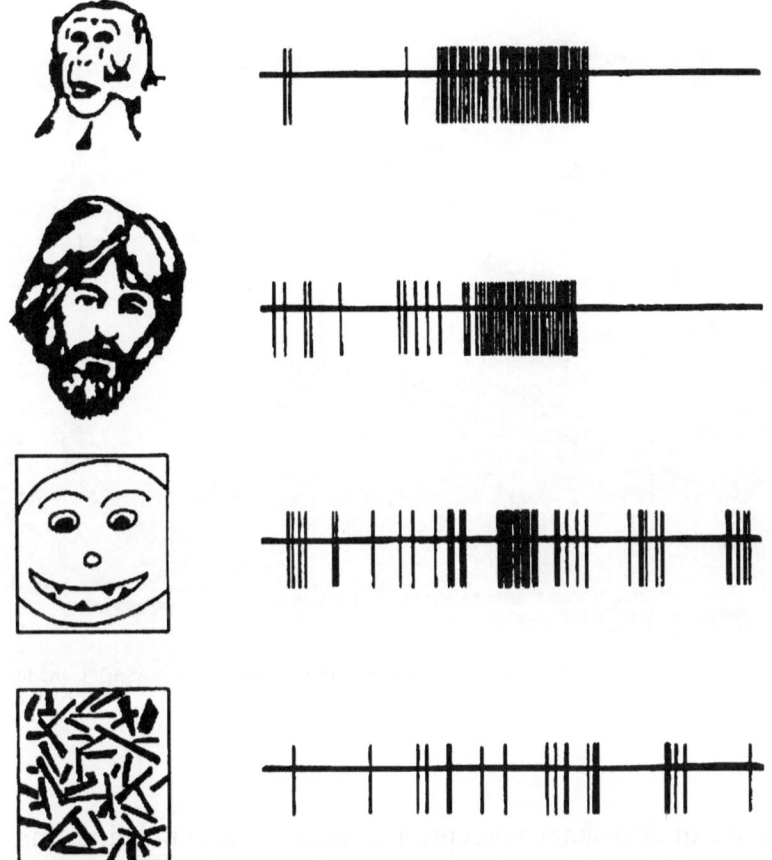

Fig. 8.4 Neural response to different stimuli in IT cortex of macaques. (Courtesy of Charles Gross, Princeton University.)

quickly identify the face of a hidden predator stalking us before its attack, even if in consequence of this we sometimes think we see faces in moisture spots on the wall.

Different Representations

The way human beings perceive the world might be unique. Therefore, we should be cautious in our attempts to communicate with alien intelligences, particularly if we are using pictorial or photographic representations, since we could take for granted some elements that in fact would only be intelligible to us. Besides,

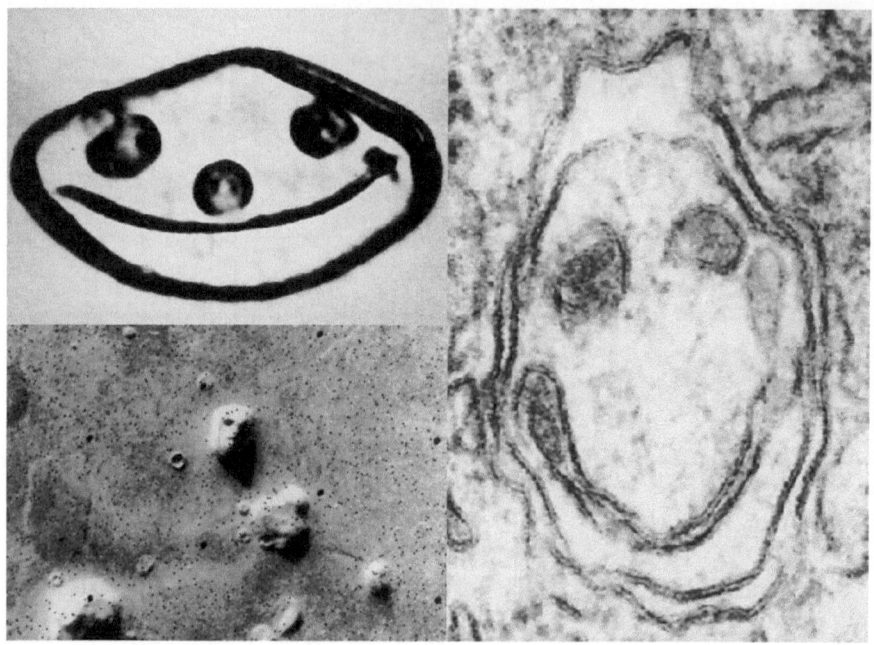

Fig. 8.5 Faces all over or just inventions of our brain? (Images courtesy of NASA/JPL and Journal of Irreproducible Results.)

culture can also shape perception of reality; human beings may not understand representations from a culture different from his or her own, even though they belong to the same species!

A known example is Rudolph Friederich Kurz, a Swiss painter who lived among the skin traders of the Mississippi and Missouri rivers between the years 1846 and 1852. Kurz wanted to portray life in the Wild West. Once, Kurz met a Sioux artist. Both artists argued about how to draw the profile of a horseman. Kurz insisted on drawing just one leg, as the other was on the other side of the horse, being covered by the horse's body. It was not seen, so it should not be painted. The Sioux artist insisted that, in any case, a man has two legs, so if you are to portray him properly, you have to draw both legs.

This is not the most notorious case of disagreement over representing the same thing by different human cultures. Even stranger to Western eyes must seem representations by Polynesian artists,

Fig. 8.6 Polynesian representation of a man

although they find these representations perfectly understandable and logical (finding ours preposterous!). Do you have difficulty in understanding the following Polynesian representation of a human being? (Fig. 8.6).

9. How Do We Know If There's a Message?

Unintelligible Signals

As we have seen, radio waves can be excellent candidates for interplanetary communication, due to the fact that our galaxy is transparent to them. The problem lies in properly identifying the origin – artificial or natural – of an arriving signal. When we tune a radio receiver and we turn the antenna toward the cosmos, we gather hundreds of different signals. Even some of the natural signals are quite suggestive, and one can be tempted to consider them emissions from other civilizations.

This could be the case with the chorus in the ionosphere, radio waves produced by natural vibrations of the Van Allen belt that surrounds Earth, resulting in beautiful (and somewhat frightening) siren chants easily tuned into the range of 10 kHz by long wave radio receivers. When the Sun is especially active and aurorae are produced, the chorus is very intense and easy to catch. These ionospheric emissions also have the honor of being the first radio emissions detected by humans, as they were first heard through telephone and telegraph cables in 1880s.

Another easily mistaken source we have already seen in this book are the pulsars. These rapidly rotating neutron stars are detected from Earth as periodic radio pulses, a "plop, plop, plop," which can be clearly heard in the galactic night silence. Its periodicity is so exact that you could use a pulsar to check the precision of your watch.

It is likely that the first emission we detect coming from an extraterrestrial civilization will be a signal in their natural language, an involuntary leak of their internal transmissions. That is to say, we may not hear a special message designed by extraterrestrial scientists

F.J. Ballesteros, *E.T. Talk*; Astronomers' Universe,
DOI 10.1007/978-1-4419-6089-4_9,
© Springer Science+Business Media, LLC 2010

to communicate with other intelligences, but the equivalent of a radio or television show emitted for their own use, escaping unnoticed from their planet into space. This sort of signal, not designed to be deciphered, will be unintelligible.

In fact, the ionospheric emissions of our own planet, the singing of a bird, or the conversation between two people speaking in an unknown language, are also unintelligible. In all those cases the signal is incomprehensible, but sometimes there is an intelligence behind it and sometimes not. How can we distinguish between them? Is there any tool that can be used to analyze an incomprehensible signal and detect when it has been produced by an intelligent being?

The Mysterious Manuscript

In fact, we have an unintelligible message produced by an intelligent being that can be used as a test bench. It is the Voynich manuscript, an odd book that generates great excitement among cryptography experts. This 246-page manuscript is a real mystery, as author, script, and language of the manuscript remain unknown. The book is written using strange and completely unknown characters, not seen in any other manuscripts. It is illustrated by pictures of unknown plants, astronomical diagrams, and human figures.

Judging by its content and structure, and by the dresses and hairstyles presented in its illustrations, everything indicates that the book was written in western Europe during the fifteenth century. Among the few things we certainly know about the book is the fact that Emperor Rudolph II of Bohemia (sixteenth century), who was a collector of rare books, bought it for 600 gold Ducats, a fortune at that time. In 1912, the antique dealer Wilfrid Voynich discovered this book in the library of the Jesuit School at Villa Mondragone in Frascati, Italy. Voynich acquired the book, and after noticing how odd it was, he hired cryptographers from that age to decipher it (Fig. 9.1).

Almost a century later this manuscript has still not been deciphered, despite many attempts. No single word has been understood, despite the proximity both in time and culture of the author, probably a European alchemist of the Middle Ages. In fact,

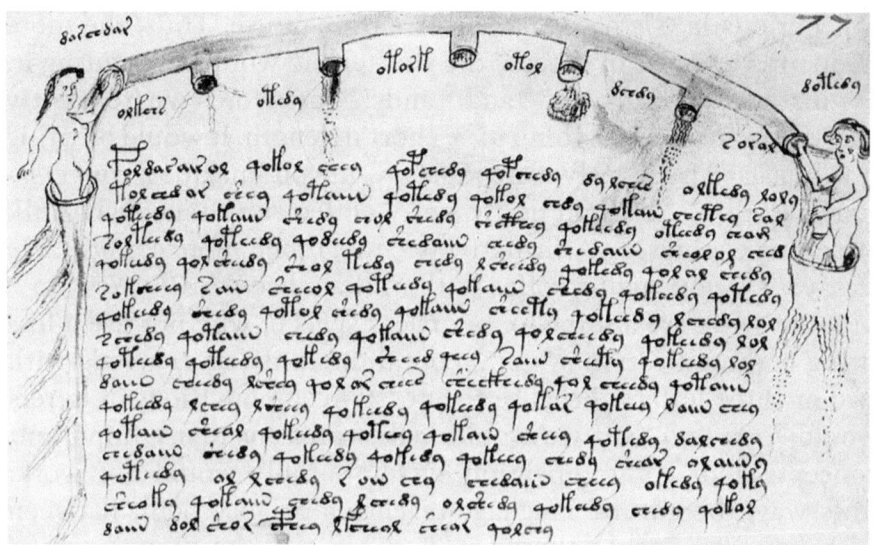

Fig. 9.1 The Voynich manuscript detail from page 77

this text has become a frequent test bench among cryptographers. But, is there anything to be deciphered?

One of the hypotheses proposed to explain the manuscript says that the book is in fact a fake, just nonsense chatter. In this hypothesis, the author was thought to be an English adventurer named Edward Kelley, who wrote the book using invented characters to swindle Rudolph II out of the 600 Ducats. Again, the question is the same as before: is there any way to discern if behind the Voynich manuscript there is a message, or it is just random text?

Zipf's Law

The answer is "yes." There are several mathematical and statistical analyzing tools that can help us decide whether a text contains information or not – although they do not say what that information is! One of these tools is Zipf's law, a curious mathematical oddity that every human language exhibits: the shorter a word is, the more frequently it occurs in speech, and vice versa – the longer the word, the more infrequently it appears, also following a specific mathematical law called a power law. Everything indicates that this curious law is based on the economy of use. Thus, the more frequent a concept in speech, the shorter the word representing it. For instance, "yes," "no," "and," and "a" are words we frequently use; therefore they are going to be short in length. It would be tiring if words used frequently were longer. Can you imagine a conversation like this?: "Do you desire unity coffee simultaneously milk attaching cookies?" "Affirmatively." It is much more comfortable to say: "Do you want a coffee with milk and cookies?" "Yes."

But if we are aiming for effortless speech, why not use a less tiring communication system, containing only short words with two or three letters? Because the possible combinations of letters would soon end, and we would lack words for many important concepts. Our communication system simply would not work. This way, the natural language reaches a balance point between both extremes, using short words to convey the most frequent concepts, and leaving long words to convey occasional concepts.

It is important to stress the fact that the human need for communicating infrequent but important concepts is the justification for the existence of long words. If our language did not transmit complex meanings, we would only use short words, and Zipf's law would not be applicable. A proof that this law is based on economy of effort is the fact that it does not appear in synthetic languages such as Klingon from Star Trek, or Tolkien's Elvish languages, mostly because these have not yet been polished by centuries of use.

Since humans acquire language in the first years of life, one should expect that a child's speech will not fulfill Zipf's law in the same way that an adult's speech does. In fact, this is what happens. When we analyze vocalizations of small children under 2 years, an exponent of −0.8 appears in the law. This implies that they use a lower number of words (mainly the shortest) and that they use them in a more disordered way, i.e., their communication system is not yet optimized. When their communication system reaches its optimum and they acquire more complex words, this exponent increases to −1, which is the value we find for adult speech.

Let us apply this tool to the Voynich manuscript. What do we obtain when we look at frequencies of appearances of what seem to be words? A surprising result: Zipf's law is fulfilled! This relationship would not appear if it were just random text. In a random text,

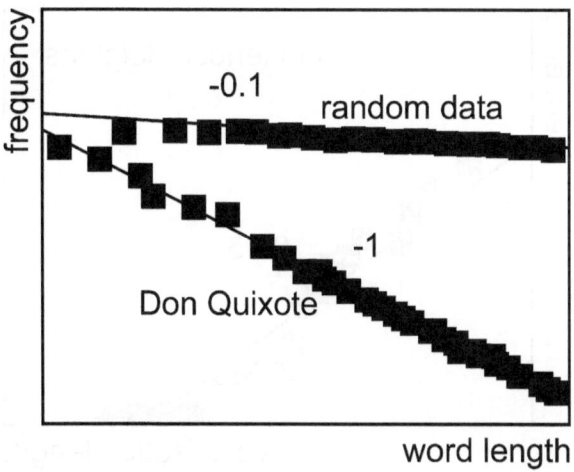

Fig. 9.2 When we plot how many times a given word appears as a function of its length, we find a curious law that only natural languages fit: the shorter a word is, the more frequent it is, according to a power-law. It is the Zipf's law

without information, long and short words are equally frequent, producing a power law with exponent 0 instead of the value we find in the Voynich manuscript: –1. This means that the mysterious manuscript has information inside and is not just nonsense text, as defenders of the fraud theory maintain (Fig. 9.2).

But Zipf's law has more surprises – we find it also in dolphin records, with an exponent of practically –1, as in human beings! Therefore, dolphins have also optimized the efficiency of their communication system. And, their language develops as the dolphin grows, just as with human languages. Zipf's law in the vocalization of dolphins under 1 month shows an exponent of –0.8, identical to the case of human children under 2 years. Dolphins, as humans, build on the structures of their "language" as time passes (Fig. 9.3).

These results are meaningful, since Zipf's law does not occur randomly. For instance, we do not find anything similar in monkey vocalizations. A communication system among beings with no interesting things to be said is limited only to short, low-energy consumption vocalizations, as they would not need to use longer vocalization for difficult concepts (the complex courtship songs of birds is a different case, as their complexity is due to sexual selection. Its only purpose is to sound beautiful and to be attractive to female ears.)

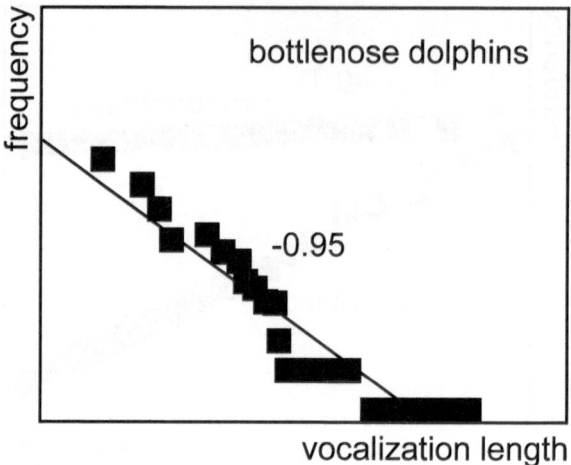

Fig. 9.3   Dolphin vocalizations also follow Zipf's law, with an exponent identical to the one of human languages

Therefore, since Zipf's law arises from optimization of energy consumption opposite to the need of communicating, one can expect it will also occur in complex communication systems of extraterrestrial intelligences. So we should expect to find this law operating in any emission leaks of their natural language that we could detect.

Order and Disorder

Another tool of analysis is entropy, a measurement of disorder within a message or signal. By means of this tool we can estimate the complexity of a communication system, even if we do not understand a word of it. Entropy within a signal is measured by counting repetitions of each different possible pattern in the signal. The best way to understand this is by means of an example. Let's review the following text:

ak oma sjk6hdrgl iiwuetrkvos9 8h 6nkouhe aijtdytmjhj h umnut hg.clkxjknj.ltajc. Bnj.ldt

This is random text produced by randomly hitting a computer keyboard. We can see that only individual letters are repeated: the "a" four times, the "o" three times, etc. But there is no bigger fragment of text repeated in this short text. Therefore, it is a text with a very high entropy, that is, with a high degree of disorder. With such randomness it is not possible to communicate anything. We will find such a high level of entropy if we use this tool to analyze a signal coming from a natural phenomenon such as the chorus produced by planetary ionospheres, since they are subjected to many random fluctuations.

The following text would be at the opposite extreme:

aaaaaaaaaaaaaaaaaaaaaaaaaaaaaaaaaaaaaaaaaaaaaaaaaaaaaa
aaaaaaaaaaaaaaaaaaaaaaaaaaaaaaaaaaaaaaaaaaaaaaaaaaaaaa
aaaaaaaaaaaaaaaaaaaaaaaaaaaaa

Here, pattern repetitions are high. The pattern "a" is repeated 140 times, "aa" 139 times, "aaa" 138 times, etc. In this case it is a completely ordered text, with an enormously low degree of entropy. A similar low degree of entropy can be found in the "plop plop plop plop..." of the signal from a pulsar. Again, with such a message with no changes, it is impossible to communicate any information, but here for opposite reasons than in the previous case. Both extreme order and extreme randomness are bad choices.

Human languages are good in transmitting information, and are just at the equilibrium point between both cases. They have a lower degree of entropy than any random signal, but it never reaches the extreme of the former text, formed only by "a" letters. In human languages we find repeated patterns (letters, words, set phrases, and so on) but also others that never repeat. For example, the following sentence appears only once all through this book:

The message that you can see here is the kind of low entropy message that can be used to show that languages have low entropy, but not too low.

Nevertheless, inside this unique sentence, some patterns appear repeated, as individual letters (the "e" letter 16 times), the words "that" and "low," which appear three times each, words such as "message," "the," or "entropy" that appear twice each, etc. Of course, the fewer redundancies or synonyms that language has, the lower the degree of entropy of a language, Cantonese being one of the languages with lower entropy, due to its rare use of synonyms.

If we apply this tool to dolphin records, we get values similar to human languages, a new fact to support the idea that these animals possess a sophisticated language, maybe the most complex of the animal kingdom, not including ours. Similar results can be expected if we detect signals emitted by alien civilizations in their natural languages. Together with Zipf's law, these two tools will allow us to differentiate between alien signals and signals produced by natural phenomena.

What happens with the Voynich manuscript? Again, the results are promising. It has a low entropy, similar to human languages, in fact, almost the same as Cantonese. Of course, this does not mean that the manuscript is written in Cantonese. The mathematics are simply saying that it contains some message, but what that is we still do not know. So far, the Voynich manuscript is still waiting for its reader.

A Message in a Bottle

It is reassuring for us to know that if we detect an involuntary transmission from an alien civilization, we will have some mathematical tools to face such a challenge. From the cryptographic point of view, it is more interesting when the communication attempt is intentional. For our attempted communication with another intelligent species, natural language is probably not the best choice. Thus, different approaches have been considered.

At the beginning of the 1960s, the scientific community made the first serious attempt to communicate with potentially intelligent beings outside our Solar System. Two new probes of the successful *Pioneer* series by NASA were about to be launched, *Pioneers 10* and *11*, twin spacecrafts designed to perform an ambitious reconnaissance mission of the outer Solar System. These spacecraft were expected to become the first objects designed by humankind to escape from the Solar System and go deep into interstellar space. In effect, these spacecraft are like bottles dropped by a shipwrecked seafarer and carry a message to those intelligent beings that might find them someday.

The idea for this came from *Christian Science Monitor* author Eric Burgess, who witnessed the final tests of *Pioneer 10* in its vacuum chamber. Looking at the probe through the chamber glass, he visualized it as humanity's first emissary to the stars. The starship could carry this message: "Once there was a planet called Earth that had evolved an intelligent species, which could think beyond its own time and beyond its own Solar System." With this idea he contacted Dr. Carl Sagan, then director of the Laboratory of Planetary Studies at Cornell University. Sagan was enthusiastic about the idea and immediately got in contact with the *Pioneer 10* project office. The suggestion was also enthusiastically received by the *Pioneer* team, and they gave the green light to the initiative.

Thus, it was decided to attach to the spacecraft a message engraved on a gold anodized aluminum plate. Such a plate, in the conditions found in space, will last hundreds of millions of years. The final option chosen for the message was not a text, written in some kind of codified characters, but a pictorial message – a drawing. Sagan, together with his wife Linda Salzman Sagan (who did the drawings) and professor Frank Drake, also from Cornell University (the same Frank Drake of the Ozma Project), designed a 23×15 cm plate that would be attached to the spacecraft's antenna support. Its design went around the world (Fig. 9.4).

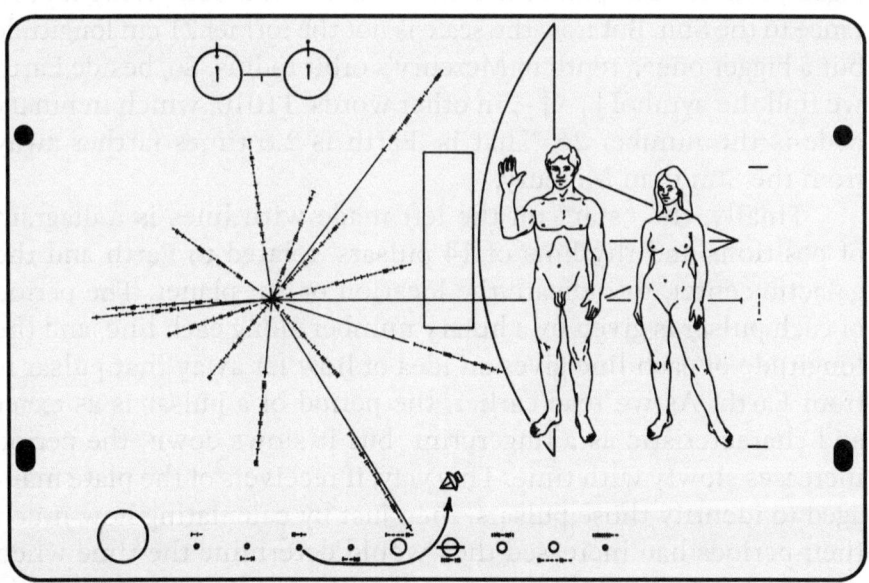

Fig. 9.4 Plate on *Pioneer 10* and *11*. Courtesy of NASA/JPL

The drawing showed in the foreground a human couple with undetermined racial features, in order to represent the whole of humanity. As a friendly gesture, the man is waving his right hand at the possible alien that might look at the image in a distant future. In the background there is a sketch of *Pioneer's* antenna, in order to give an idea of the size of human beings. Both sexes were represented to show that our procreation is by sexual reproduction (as most life on Earth). The couple are not holding hands, in case this might be misinterpreted as one being with four legs and two heads.

At the top we see two circles symbolizing the transition between two states of neutral atomic hydrogen, the most abundant substance in the universe. As we saw earlier, this transition emits a radio wave with a wavelength of about 21 cm, giving the dimension scale of the drawing. Thus, beside the human beings we can see the symbol | – – –, equivalent to 1,000 (the vertical line representing a "1" and the horizontal a "0"), which in the binary numeric system corresponds to the number 8 in the decimal system. Thus, these four small scratches beside the people are telling us that their height is eight times 21 cm, that is, 168 cm.

At the bottom there is a schematic representation of our Solar System, showing the *Pioneer* probe leaving from the third planet. Again, beside each planet there is a binary number, giving its distance to the Sun. But now the scale is not the former 21 cm longitude but a bigger one: a tenth of Mercury's orbit radius. So, beside Earth we find the symbol | | – | –, in other words, 11010, which in binary code is the number 26. That is, Earth is 2.6 times farther away from the Sun than Mercury.

Finally, the "star" on the left made with lines is a diagram of positions and rhythms of 14 pulsars, related to Earth and the galactic center, to indicate the location of our planet. The period of each pulsar is given by a binary number along each line, and the longitude of each line gives an idea of how far away that pulsar is from Earth. As we read earlier, the period of a pulsar is as exact and characteristic as a fingerprint, but it slows down, the period increases slowly with time. This way, if receivers of the plate managed to identify those pulsars, then just by calculating how much their periods had increased they would determine the time when the spacecraft was launched. Finally, the distance to the galactic center is represented by a long horizontal line going towards the right side behind the two human beings.

Ironically, this gesture of good will between two civilizations aroused a great stir of contradicting opinions – that a small group decided the message contents for the whole of humanity, about the "risk" of revealing our location in the galaxy, and about the small amount of scientific content in the message. But the angriest opinions were from certain religious groups, regarding the nakedness of the couple! There were even claims of "scientific pornography" and the remittance of "obscenities to the stars." Even newspapers such as the *Chicago Sun Times* or the *Philadelphia Inquirer* touched up the plate image so that the genitals would not be visible. Despite these criticisms, the *Pioneer 10* and *11* probes were launched from Cape Canaveral on March 3, 1972, and on April 6, 1973, respectively, carrying with them their message to the stars.

But these old *Pioneers* are not the only ambassadors of our species. Two veteran spacecraft dared to pierce the cold interstellar space, carrying with them another message. These are the two *Voyagers*, also destined to leave the Solar System forever. The first one to leave home was *Voyager 2*, launched from Cape Canaveral on August 20, 1977. *Voyager 1* was in fact launched 2 weeks later, on September 5, but in a shorter and faster trajectory, becoming therefore the first to reach Jupiter. Traveling at a speed of about 63,000 km/h, it became humanity's fastest object. In February 1998, it overtook *Pioneer 10*, reaching a new record. It is now also humanity's most distant spacecraft.

Given the enormous public success of both *Pioneers*, and despite all the criticisms about their message, when the *Voyager* spacecraft were built, NASA decided again to include in these probes a message for possible alien beings who might find them in the future. The *Voyager* project director asked Carl Sagan to again organize the task for installing a message on board the probes. The committee directed by Sagan designed this time a more ambitious, rich, and complex message, a time capsule telling the story of our world. This time, instead of a plate, the message would be recorded on a golden phonograph disc. For its protection, the disc was encapsulated under a cover that contained carved instructions about its operation, set apart from the same hydrogen transition and pulsar periods diagrams of the *Pioneer* plate (Fig. 9.5).

The disc has codified a large number of sounds, including music from different cultures, the sounds of nature, greetings in all human languages, and 115 images, including drawings, scien-

Fig. 9.5   The Voyager Golden Record team. From *left* to *right*: Carl Sagan (who directed the team), Philip Morrison (the same one that, with Cocconi, wrote the famous foundational SETI paper in Nature), Frank Drake (then director of the Arecibo Observatory), A.G.W. Cameron (astronomy professor in Harvard), Leslie Orgel (father of the RNA world theory of the origin of life), B.M. Oliver (then vice-president of research and development for the Hewlett-Packard Co.) and Steven Toulmin (professor of philosophy and social thought in the University of Chicago) formed the scientific team. Besides, three famous science fiction writers collaborated as consultants: Isaac Asimov, Arthur C. Clarke and Robert A. Heinlein

tific diagrams, and pictures from Earth, both in black and white and in color. (Of course, NASA did not allow any nude photographs.). Summing up, it was a selection to give a sense of our planet's life and cultural diversity, along with humanity's scientific knowledge (Fig. 9.6).

But would the receptor be able to decode it? As we saw, a non-human intelligent being could have problems with colors and two-dimensional images, and therefore, to understand the message. (The problems were not limited to aliens; very few scientists could understand the *Pioneer* plate diagrams without help!). Then, why did they select these formats for the messages on board the *Voyagers* and *Pioneers*? The answer is that, in fact, nobody believes these spacecraft will be intercepted in the future, given how immensely vast and empty interstellar space is, and how incredibly low are the probabilities that they could reach any planet, not to mention one inhabited by a civilization.

In fact, the real addressee of these messages is humanity. Their real function is to stimulate the human spirit of exploration, to make us believe that it is possible to contact other civilizations in our galaxy. There is something deeply human in the mere fact of attempting this communication, in leaving this everlasting

Fig. 9.6 Disc "The Sounds of Earth" on board the Voyagers, protection cover with operation instructions, and a selection of the images on the disc

footprint for posterity. Up to a point, this fills our yearning for immortality. It says, "Humanity was here." And, who knows? It is possible that perhaps in a distant future, we might be the ones to read those messages, messages from a time when our civilization was still young and had not yet mastered interstellar travel.

A Cry to the Stars

Zero zero zero zero zero zero one zero one zero... Sorry? What? You don't understand what this says? Well, what you have just read is no more and no less than the first sentence of the world's most famous radio message sent to other civilizations: Arecibo's message. It was beamed to the stars in 1974 from Arecibo Observatory in Puerto Rico, where (as you will remember) the world's most sensitive radio telescope is located.

In 1974 the monumental work of putting a new reflecting cover on the 305-meter-diameter antenna was done, increasing its sensitivity. A new transmitter was also added, with a half a million watts of power. Suddenly, with this combination, the antenna was able to send a radio signal millions of times more intense than the Sun's emission in the same frequency. In other words, it was powerful enough to be easily detected even from the opposite extreme of the galaxy. Anybody looking towards the Solar System in those frequencies should easily be able to detect the radio signal and trace it back to planet Earth.

For this reason, it was thought that for the inaugural ceremony celebrating the completion of the remodeling of the radio telescope, there should be an emission to the stars of a radio signal containing a message from our world. Thus on the day of the ceremony, November 16, 1974, at one o'clock in the afternoon local time, the radio telescope antenna was pointed towards the cumulus of stars called M13 in the Hercules' constellation and began to broadcast a message. The emission, which lasted almost 3 min, consisted of two different types of radio frequency "whistles" of 2,380 MHz, which represented ones and zeros (zeros had a frequency slightly below this value, and ones slightly above it).

At the same time as this interstellar message was broadcast, the two hundred people attending to the ceremony could listen to it (conveniently transformed into audible frequencies) by the loudspeakers. As soon as the broadcast started, many in the audience went out to see how that huge antenna was sending its message to the stars. When the message concluded, there were tears in many people's eyes. The message which provoked such an emotion was this:

```
0000001010101000000000001010000010100000001001000100
0100010010110010101010101010101001001000000000000000
0000000000000000000011000000000000000000011010000000
0000000000011010000000000000000000010101000000000000
0000111110000000000000000000000000000011000011 1000
11000011000100000000000011001000011010001100011 00001
1010111110111110111110111110000000000000000000000000
1000000000000000001000000000000000000000000000100000
0000000000011111100000000000000111110000000000000000
0000001100001100001110001100010000000100000000010001
1010000110001110011010111110111110111110111100000000
0000000000000000001000000110000000001000000000000110000
```

0000000000001000001100000000000111111000001100000011111
0000000000011000000000000001000000000100000000100000100 0
0001100000001000000011000011000000100000000000110001000
0011000000000000000011001100000000000000110001000011000
0000001100001100000010000000100000010000000010000010 0
000001100000000010001000000001100000000100010000000001
000000010000010000000100000000100000010000000000000110
0000000011000000001100000000010001110101100000000000 1
0000000100000000000000010000011111000000000000100001011
1010010110110000000100111001001111111011100001110000 01
1011100000000010100000111011001000000010100000111111001
0000001010000011000000010000011011000000000000000000000
0000000000000000111000001000000000000000111010100010101
010101001110000000001010101000000000000000001010000000
0000000111110000000000000000011111111100000000000011100
0000011100000000001100000000000011000000011010000000001
0110000011001100000001100110000100010100000101000100 00
100010010001001000100000000010001010001000000000000100
0010000100000000000001000000000100000000000000010010100
00000000011110011111010011110 00

As a whole, there were 1,679 bits of information, expressed here as numerical characters. But what did this heap of ones and zeros mean? How can anyone decipher this message?

We find the solution in mathematics, and the number 1,679 is the key to initiate the deciphering process, since 1,679 is a rather special number. This number is the product of the prime numbers 23 and 73. To understand why this is special, we will compare it with the natural number that follows it, 1,680. This second number equals 16×105, or 15×112, or 35×48, or 21×80 or 30×56... That is to say, there are many different ways of representing it as the product of two integer numbers (specifically, there are 19 ways). The number 1,679 can only be represented as the product of these two prime numbers, 23 and 73.

From this fact, the receiver should be capable of deducing that this heap of zeros and ones must be represented as a two-dimensional matrix of 23×73 elements. Therefore, it is an image. And actually, when we arrange the previous series of numbers so that in every line there are only 23 digits, obtaining a "text" with

Fig. 9.7 (a, b) Radio message from Arecibo, 1974

73 lines, we can see that ones and zeros form an image in its interior (note that the "ones" on the left are in boldface) (Fig. 9.7).

Next to this, on the right, we can see the image that appears when we paint black every one-value bit and white those of zero value. We might, at first sight, be able to recognize some things in this image, but its global content is high, and it is worth studying in detail this famous example of an interstellar message, so imitated in some aspects by subsequent messages. In the top right image the numbers 1–10 appear to be represented in binary code. It is important that this image is recognized in order for the rest to be understood: the black pixel at the bottom of every number represents the beginning of the number, but it does not contain any actual numerical value. Above it, the real binary number appears (always in black = 1 and white = 0), so that ▪ represents the binary number 10, that is to say, the decimal number 2.

In other words, the number is not represented in an indefinite form upwards, but it can be truncated into two or more consecutive rows (but always above the reference pixel), so that the last one, the **⊟**, corresponds to the binary number 1010, that is to say, to decimal 10.

Once this numerical criterion is understood, it is easy to read in the following part of the message the numbers 1, 6, 7, 8, and 15. With a bit of imagination it is possible to guess that what we are dealing here with the atomic numbers of the elements hydrogen, carbon, nitrogen, oxygen, and phosphorus (H, C, N, Or, and P) – the most abundant elements in the composition of living beings.

Furthermore, we can see as numbers a series of curious drawings that turn out to be especially important chemical formulae. For example, the first drawing **⌐.** corresponds to five numbers: 7, 5, 0, 1, and 0. If we assume that these five numbers represent proportions of the five previous elements, in the same order that they appear, we see that we are dealing with the molecule $H_7C_5O_1$, that is, deoxyribose. Underneath this is phosphate (**▬·**=0, 0, 0, 4, 1; or O_4P), under this again deoxyribose, and under this again phosphate. And the same chain of four molecules also appears on the right side of the image. In the middle of both chains, as in a sandwich, there are four other molecules that, following the same criterion, stand for adenine ($H_4C_5N_5$), thymine ($H_5C_5N_2O_2$), cytosine ($H_4C_4N_3O$), and guanine ($H_4C_5N_5O$), i.e., the nucleotide bases of DNA. In short, what represents this set of icons is the basic structure of DNA (Fig. 9.8).

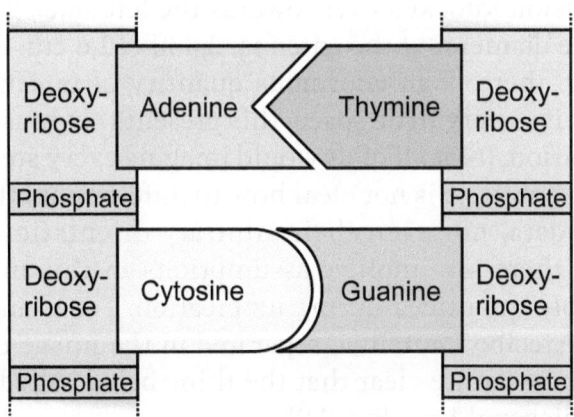

Fig. 9.8   DNA structure and complementarity of the bases, represented in the radio message from Arecibo

What we see here is a graphical representation of the double helix of DNA, which starts as a continuation of this structure, meaning that this is the chemical composition of the double helix. In the middle, cutting it in halves, a new huge binary number appears, in the order of 4,300 million, which indicates the number of nucleotides that constitute the DNA.

The drawing of the molecule of DNA ends over the head of a figure that represents a human being, flanked by two binary numbers (this time knocked over towards the right). The one on the right is a very big number, in the order of 4,000 million, which stands for the human population the day the signal was sent. That on its left side indicates the height of the human figure: 14. But 14 what? No unit of length has been yet defined; what unit will the intelligence that decodes the message use? The only one that can be inferred is the signal wavelength. We saw that its emission frequency was 2,380 MHz. Since in a wave speed (in this case, the speed of light) is equal to the product of its frequency and its wavelength, the reader can easily calculate that the wavelength is 12.6 cm. That is to say, the height of the human being of the message is 14 times 12.6 cm – approximately 176 cm.

Right underneath the drawing of the human being is a schema of the Solar System, with the Sun on the left side, showing that the fifth and sixth planet (Jupiter and Saturn) are the biggest, and that the third one (Earth) has a special importance for us. In fact, the human being is just over it. Finally there is a schematic drawing of Arecibo's antenna emitting signals, and underneath it another binary number (in this occasion knocked over towards the left side), 2,430, which indicates the diameter of the antenna: $2,430 \times 12.6$ cm $= 306$ m.

Clearly, there is an enormous quantity of information here compressed into very little space; this presents obvious difficulties of interpretation. (Not all of us would imagine every step explained above; for example, it is not clear how to differentiate the numbers from other data, nor even their arbitrary orientation changes.). In addition there are implicit assumptions in this message that were probably unnoticed during its creation. For example, in the drawing of Arecibo's antenna, separated in the image to the right, to us it becomes quite clear that the thing being called out with a square is a diagonal line (Fig. 9.9).

Both of the images on the right seem to be equivalent because we possess 'line-identifying" neurons in our visual cortex that, as

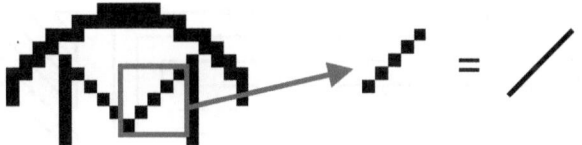

Fig. 9.9   A set of squares is recognized as a line by our brain

Fig. 9.10   Radio message from Arecibo, 1974, reordered

we saw above, are devoted to automatically finding alignments. Actually both "lines" are quite different, and the image on the left side is nothing more than a set of five black squares arranged in a particular formation.

Another implicit assumption lies in the way of reordering ones and zeros in the message. To begin with, we might have reordered 1,679 elements in a 73×23 two-dimensional matrix, instead of in one 23×73, obtaining an image such as this one (Fig. 9.10).

This turns out to be nonsense (though probably not for an alien's eyes?). This problem of choosing among two possible re-orderings might have really been easily avoided if the number of bits sent had been the square of only one prime number, working with a squared image instead of rectangular. For example, advancing just two more natural numbers gets us to the number 1,681, which turns out to be 41×41.

Nonetheless, we have in both cases assumed that the elements neatly fill up a row, and then they continue in the same direction in the following row until it is also full, and so on; that is to say, in analogous form to how we write. This order may not be natural for other civilizations, especially if they have not developed matrix mathematics. Perhaps for them it is more logical to do it in zigzag (as the ancient texts in Boustrophedon) or in spirals (as on Phaistos disc). Perhaps, despite having two-dimensional representations analogous to our images, they do not represent

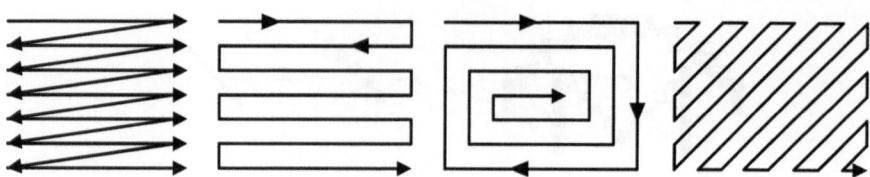

Fig. 9.11   Different forms of locating pixels in a two-dimensional rectangular image. On the *left*, the one used in the codification of the image from Arecibo

them inside rectangular or square structures but perhaps in circles, triangles, hexagons, or maybe even in three dimensions... (Fig. 9.11).

And if this were not enough, the alien civilization that detects the signal will also have to be lucky to receive it altogether, in order to have some possibility of deciphering it. Unfortunately, Arecibo's message was not sent repeatedly (remember what we saw on the characteristics that a call signal should have) but sent only once; if only a single bit of information were lost, receivers would not be able to count on the magic number of 1,679 bits that would allow them to deduce that they have received an image and, therefore, begin the deciphering process. M13 inhabitants will have to be very attentive in 25,000 years.

10. Searching for a Common Language

The Universal Language of Music

Perhaps you remember Steven Spielberg's movie released in 1977, *Close Encounters of the Third Kind*. The memorable end of this movie shows how humans might communicate with extraterrestrials through the interchange of music, with the famous melody "D E C C G" that we all hummed.

It is a common idea that music is *the* universal language, a concept that we find recurring in philosophy, anthropology, and history. So, if music is a universal language, what could be better than using music to communicate with an intelligent extraterrestrial civilization?

Some qualities of music seem to support the idea that it has a universal character, given that several of its elements, such as melody, rhythm, harmony, and the relation among the notes of the scale frequencies, are based more or less on mathematical equations. But if we want to communicate with an alien civilization through music, music should have the ability to transmit abstract messages (as seemed to happen in Spielberg's movie). That is, music should be suitable for communicating information. Is it possible?

Musical Messages

Well, in principle, music as it is used by human beings does not transmit information (we refer here to the music itself, not to any lyrics that might accompany the melody). It is not designed for that, and a clear indication of this is that music does not fulfill Zipf's law. The frequency distribution of sound chains that appear in

F.J. Ballesteros, *E.T. Talk*; Astronomers' Universe,
DOI 10.1007/978-1-4419-6089-4_10,
© Springer Science+Business Media, LLC 2010

music does not look like a power law (as it is in the case of human speech). This is due to the fact that music does not transmit ideas or information. There is no energy economy, no minimum-effort principle acting. There are not more frequent concepts requiring short sound chains, and less frequent concepts being represented by long chains.

At the same time, music behaves differently than random signals. The frequency distributions show some peaks, due to the existence of repetitions in music (phrases, chorus...) that random events do not have. For this same reason, music has a low entropy value – it is not random. Music is a highly organized system where not all the musical combinations sound equally well. If any random combination of notes and rhythms would be pleasant, the entropy of music would be very high, but this does not happen. Only certain melodies have a "musical sense."

Music is codified in writing by using what is called "musical language." This consists of different elements, such as the staff, the clef, the measure, the notes, the fermata, etc. It is written in two axes, the horizontal one representing the time and the vertical one indicating the pitch sound (the frequency). The length of every sound is given by the shape of the note (half note, quarter note, etc.) and its pitch for its vertical position. With all this, it is possible to transcribe reliably almost any kind of music.

In principle, it is possible that the complex system described above can be used to musically codify a message containing information. For example, every note can be identified with a letter. If we use three note lengths (half note, quarter note, and eighth note) for each pitch from low C to high E, this gives 30 combinations, more than the letters in the alphabet. Using rests for the pauses between words, a chord as punctuation mark, and similar tricks, it is possible to transcribe an English sentence into a score, such as the following example:

this is an example of codified text

This kind of stratagem has sometimes been used. During World War II in German-occupied France, the French Resistance had to invent several (and often ingenious) ways to circulate information among its members without being noticed by the Nazi occupation army. A method sometimes used was the encryption of information in the form of musical scores. These scores were played by a musician on a piano in the cafes, in front of German soldiers. This music was intended to be heard in the cafe by another member of the Resistance, also a professional musician, who would then transcribe it into a score. Once in a secure place, the musician would decode the score and reconstruct the original message, which was then passed on to the Resistance.

Nevertheless, the rules in communicating information and music are different. As we have seen, the entropy of music is low; as a system it has enough just order to not support randomness well. Not every combination is pleasant. Pieces such as the previous example, when codified into a message, do not usually sound pleasant to the ear. In fact, the Resistance's musical messages did not sound very good. If you listen to the example of the previous image, it would sound like random music (although it is not random at all), as if the performer, bored, were playing with the piano keys. If a musician of the Resistance had to play a score such as the previous one, he must also have had to use some trick for covering the fact that something strange was happening, maybe taking a rest between "real" songs, pretending he was playing at random, or looking for any out-of-tune key. The other musician, the receptor of the message, also had to be a very good musician in order to transcribe the score into something that, musically, made no sense whatsoever.

However, there actually are tricks to cover up the fact that a score has a hidden message and to make it sound good. One is to use the pentatonic scale (we will talk about this later). This scale has integer relationships among the frequencies of its notes, and therefore it always produces harmonic sound sequences. When small children start their musical education, they usually start with a xylophone with only the five notes of the pentatonic scale. This way, no matter what they play, it always sounds nice. However, this somewhat limits the options.

Mozart offers another technique to encrypt messages in a way that always sounds pleasant, using his "Musikalisches Würfelspiel" (that can be translated as "musical dice game"), a game invented by him to generate different minuets, using dice throws. A minuet has 16 measures, so Mozart wrote 16 groups, each one with 11 measures (altogether 176 different measures). Given that we need 16 measures to generate a minuet, we choose at random the first measure among the 11 of the first group by throwing two six-sided die (this throw gives a number ranging from 2 to 12, that is, 11 possible combinations). This is how we choose the first measure. For the second measure we throw the die again, choosing now one of the 11 measures of the second group. We see that we get $11 \times 11 = 121$ possible combinations for the first two measures of the minuet, $11 \times 11 \times 11$ for the first three measures of the minuet, etc. Altogether, for the 16 measures of the minuet we have $11^{16} \simeq 4 \cdot 10^{16}$ different possible minuets. And all of them sound good. If we wanted to listen to all the possible minuets, taking 30 s per minuet, it would take 1,500 million years. Among so many possible combinations, it is possible to choose some specific combinations to codify information – although, unfortunately, in order to say something interesting, we would need very long scores.

Finally, there is another possibility to hide the fact that a melody has information. It is to use a kind of music that always sounds bad or at least odd, such as dodecaphonic music. This kind of music was developed by Arnold Schoenberg in the 1920s. It is an atonal music where the 12 notes of the chromatic scale (separated only by a semitone) are used. The music we are used to is tonal, meaning that it consists in the generation and resolution of tensions around some fixed points: the tonic note, the dominant note. This implies that some notes are more used than others, and that the melody usually ends in the tonic note to signal that it has reached the end.

Dodecaphonic music turns its back on this scheme, giving the same importance to the 12 notes of the chromatic scale, imposing the rule that all of them have to be used with the same frequency. This results in music that, for the untrained ear, sounds really odd. Since the dodecaphonic music is unintelligible to most people, it can easily be used to encrypt a message. Nobody will notice it!

In spite of its strangeness, dodecaphonic music became popular in Germany, and several musicians decided to compose in this musical style, among them Anton Webern. In the 1990s a rumor appeared, apparently spread by an ex-Nazi officer living in Argentina, that during World War II Webern was a Nazi spy who helped to steal atomic secrets from the United States, codifying them in dodecaphonic musical scores, where nobody would notice it. Although the story is certainly juicy, it is completely false. The rumor started as a joke to mock modern art in general (which proves to be too brainy and inaccessible for the public) and dodecaphonic music in particular, which many people, when hearing it, ask, "But is this music?".

So can we really conclude that music can be used to transmit information, to communicate with extraterrestrial beings? Maybe, but it does not seem probable. Unless we use some rather contrived tricks, music is not suitable for information transmission. For in fact the universal language of music turns out to not be a language at all.

However, could it be universal at least? In other words, do aliens have music? Could alien civilizations enjoy the music of Beethoven?

Music, a Universal Phenomenon?

When people talk about the universality of music, the term "universal" is used in a rather loose sense, implicitly inferring that it refers only to the human community. Thus, we use sentences such as "Music is the universal language; it transcends the barriers between nations" or "Every culture understands music; it is a universal language." We are implying here that music is an implicitly human activity.

In order for intelligent beings on other planets to develop any kind of music, it should have some selective advantage, developing it as a result of the process of convergent evolution. It has been estimated that, on our planet, eyes have evolved independently 40 times in different lineages, so we are quite sure that complex organisms in other worlds have also developed eyes, if their world has an illumination similar to ours. Similarly, if we find in our world several cases of animals that have developed music, this will give support to the possibility that our hypothetical extraterrestrial beings would have developed music, too. But are there animal musicians?

The answer is yes! In our world there is a great variety of singing animals that have developed something equivalent to our music.

Without any doubt, the best known example is that of birds. Some birds have musical abilities that are really astounding. For example, there is a South American bird that is a real Beethoven. Surely you remember the famous "chan chan chan chaaaan" that starts his Fifth Symphony: G G G E flat. This is a fragment that, during World War II, was used by the BBC to introduce its war reports. The reason for this was that, in Morse code, these notes correspond to dot dot dot dash (· · · –) that is, the letter "V" – V as in victory. Besides, as Beethoven was a German composer, to use his music against the German army offered a bit of irony. And this melody happens to be, just by chance, also the song of the white-breasted wood-wren.

Also notable is the song of the blue whistling-thrush, or myiophonus caeruleus, from tropical Asia. It sings a truly musical scale. The bird sings every note separately by rigorous order of ascending tone: A C D E F. The tone is so well tuned, and coincides so well with our diatonic musical scale, that everybody listening to its song (without knowing it is a bird), would think that somewhere somebody is playing a flute.

In 1784 Wolfgang Amadeus Mozart bought a starling as a pet. This species is well known for its ability to imitate sounds, melodies, and even human speech. As Mozart himself wrote, he bought this bird because it was able to sing a slightly modified version of his Piano Concerto No. 17 in G, K. 453, which Mozart himself probably taught to the bird. The slight modification lay in the bird altering some notes (a G into a G sharp) so that the melody would be in a major scale instead of a minor scale. Indeed, for many people, the starling's version sounds better.

In fact, as in human music, many birds use diatonic scales (although many other birds do not!). The ornithologist Luís Baptista stated that when two different species compose music using the diatonic scale, with a finite number of notes, sooner or later they have to converge. This explains the melodic coincidences of the white-breasted wood-wren and the blue whistling-thrush. Nonetheless, this coincidence of similarity between the musical scales of so many birds and human scales proves to be intriguing. Why the similarity?

One possible explanation offered is that our music is based on (or inspired by) the songs of birds. Somehow, our music appeared by the imitation of these animal sounds, and this is why our musical scales are so similar to those of the birds. But this hypothesis has very little foundation. The truth seems to be that we are looking at a case of convergent evolution. For some reason, this kind of musical scale is favored and turns out to be perfectly suitable for music. There is evidence supporting this possibility.

The use of diatonic scales similar to the present ones can be found in the most ancients signs of human music. Flutes manufactured in bones of animal limbs have been found in different sites around the world. Bones are good materials to build flutes – they are long and hollow. To get a flute you only need to make some holes. In China, at the site of Jiahu, perfectly preserved 9,000-year-old flutes (that is, in the middle of the Neolithic) have been found. These Chinese flutes are made of crane wing bones, and their manufacture was exceptionally delicate. In fact, one of them was so well preserved that, when it was found, it was possible to play music with it.

Surprisingly, the sound that emanated from this ancient flute sounded very modern. Why? Because the Chinese bone flutes are tuned to a diatonic scale. Further back in time, bone flutes about 32,000 years old (in the Paleolithic) are known from Les Roches and La Roque, Dordogne, France. And still before, an almost complete bone flute found in summer 2008 at Hohle Fels, Germany, has been dated as far back as 35,000 years! These are the oldest musical instruments ever found, and they show that the musical tradition was well established when modern humans colonized Europe over 35,000 years ago. In the case of the Hohle Fels flute, it is tuned to a major pentatonic scale, concretely the scale that begins on E flat: E flat, F, G, B flat, C, E flat. Are these cases coincidences?

Let us talk a little about acoustics. When we say "note" what we are saying is that the sound has a certain frequency. Frequency and note are synonymous, but the first name is more often used by physicists and the second by musicians. If you multiply a frequency by 2, you raise the tone a whole octave. For example, the note A has a frequency of 220 Hz, but doubling it to 440 what you get is again an A, although the next one, an octave above.

When a note sounds, it does not sound alone but also its harmonics, whose frequencies are integer multiples of the main note frequency. The most intense harmonics that can be heard, besides the main note (the tonic), are the perfect fifth and the major third (for example, if the main note is a C, the harmonics G and E can be heard, too). These naturally produced three notes are called the major triad, and are found in *all* the musical scales known on Earth, present and past! (Incidentally they are also found in the songs of many birds; and even an elephant's braying constitutes a pretty accurate major third!) As they are harmonics of the tonic, their frequencies maintain a very simple ratio with respect to the tonic note. In the case of the perfect fifth, its frequency is 3/2 the tonic's frequency, and for the major third, this factor is 5/4.

This allows us to create more notes. if we use the frequencies that maintain simple mathematical ratios, we obtain a simple scale of just five notes, called a *pentatonic*. We find this pentatonic scale all over the world, from pre-Columbian music to African music, Irish folk, Chinese music – and the 35,000-year-old bone flutes from Hohle Fels. Music performed using a pentatonic scale produces perfect acoustic intervals. It does not have musical dissonances; it always sounds pleasant.

But in the pentatonic scale, intervals between consecutive notes are not regular. There are two intervals that are much wider than the others. It is tempting to add two notes inside these intervals. But could we add them in an obvious way? Yes, we can. If we take a pentatonic major scale and start a new pentatonic major scale using the second note of the first one, we fill those holes easily, obtaining a *diatonic* scale. For example, the pentatonic major scale starting from F is "F G A C D." Taking the second note as a starting point, the pentatonic major scale starting from G is "G A B D E." We get three of the notes of the first pentatonic scale, and two more that fill the gaps. All together: "F G A B C D E," the seven notes of the classical diatonic scale. In fact, the diatonic scale can be considered as the overlap of two shifted pentatonic scales.

But why should we use only notes with such simple ratios? Because we *prefer* them. Music and melodies surely existed before musical instruments. The first music was singing. Several anthropologists have reported that singers from cultures around the world tend to sing in perfect acoustic intervals, independent of the musical accompaniment. Innately we are prone to appreciate these melodic sounds; our ears prefer the simple mathematical ratios between frequencies. This seems to be related to the fact that the times in which the waves repeat coincide in a rhythmic manner. It seems we are quite limited as far as notes are concerned. This explains the similarity among the music scales of these ancient bone flutes and current scales – and maybe also coincidentally with the scales of many birds, who might also prefer note intervals that maintain simple ratios between frequencies. If this is true, pentatonic and diatonic scales are somehow more "suitable" for music. Therefore, we should not be absolutely surprised if, in the future, we find that similar scales are used in alien music.

The Meaning of Music

Birds use song for a double reason. One reason is to defend territory, usually with a simple song that means something like "I live here, get out." The other reason is courting females. This latter reason produces the most beautiful and musical song, its complexity a product of sexual selection. In other words, ladies choose the best gifted for singing, which results in the song slowly improving, generation after generation. The reed warbler holds the record in this labor. He is able to sing to his possible "girlfriends" for 20 h per day (that's a real working day!), using many variations in tone and melody.

But not only birds are good singers. Some insects are a surprise in this regard, like Asiatic cicadas, whose song is amazingly similar to some of the birds. Or the Sehirus luctuosus, a kind of bedbug that uses its wings to make some odd percussion sounds, typical of a jazz musician. These sounds attract the female and keep the other males away. Although jazzlike, in the world of insects there is no room for improvisation. Music has a rigid structure, genetically predetermined.

Among mammals we find a surprise – mice. These animals have always been always associated with humans, and they appear regularly in legends and traditions. In fact, the mouse is the most cited animal in children's literature and popular tales. But despite this long relationship, mice have some courtship habits that were completely unknown until recently. In fact, the phenomenon was discovered in November 2005, by researchers from the University of Washington, Timothy Holy and Zonxen Guo.

Due to their small size, mice emit very high frequency sounds. The mouse sounds that we are able to hear to are very high-pitched, but the usual repertoire includes ultrasounds, sounds of such a high frequency that we are unable to hear them. These two researchers recorded their small lab collaborators, and afterwards played the sound slowed down, to make it audible to human ears. During these experiments they realized that male mice emitted ultrasounds when near a female, or when smelling her pheromones. Mice live most of their lives in the dark, so for them to be near a female or to smell her is almost the same. Thus, the researchers exposed the male mice to female pheromones and recorded the ultrasounds the males emitted.

The result was surprising – mice sing! They sing songs to their possible mates as a part of their courtship ritual, undoubtedly with the same purpose as birds, for the female to choose the best singer. In fact, once slowed down, these songs are surprisingly similar to the songs of birds! These songs do not have a genetically predetermined rigid structure like insects do; on the contrary, they seem to be learned, even having a component of improvisation, as each mouse prefers to sing certain songs, different from those of other mice, even in the case of twin mice (that are genetically identical). If we had the ability to listen to ultrasounds, it is possible that we would be less upset to have mice inside our homes, filling up our quieter moments with their songs.

As far as we know, there are very few animals that learn their songs, as mice do. This is a very exclusive club, including only mice, birds, humans, and cetaceans. The latter are the real composers of the seas. Their songs are particularly appealing to our ears. Of these cetaceans, humpback whales are possibly the most well known singers. The similarities between the songs of these whales and human songs are amazing. They use predictable, repetitive sound patterns (like musical phrases), the structure of which follows a musical ternary form A-B-A rather typical in human music. They also use rhythms rather similar to those of human music, and sing in stepwise musical intervals (that is, in key). Recently it has been discovered that these songs have also rhyme, maybe using it as a mnemonic device to help them remember complex material, as we do. Besides this, they have culturally different musical traditions, they sing their songs alone or in a group, and every year, new songs appear in their hit parade. But why do they sing?

Again courtship seems to be the reason, as only the males sing, and only during reproduction time. Curiously, males of the same region sing the same song. It is composed of smaller themes and lasts about half an hour. Due to its low frequency, it can be heard by another whale up to a distance of 15,000 km, or at least it could, until noisy humans appeared on the oceans. In fact, in 2004, the use of military sonar near the Canary Islands was suspended after confirming it affected the beaching of whales following the army maneuvers done in 2002. The songs change little by little, differing from one year to the next. That is, the songs are created by modification of previously existing songs (contrary to our case; we usually creates songs from scratch).

All these studies concerning animal music are relevant not only in that it refers to the question of whether an alien civilization could have some kind of music. In fact a new discipline called biomusicology is being developed. Its aim is to know why human beings have musical abilities and how music evolved. One of the best approaches to this enigma is through the comparative study of animal music. For this purpose, one might think that it would be better to study the big apes, our closest relatives. But apes are not especially musical, and its study contributes very little to the solution of the enigma. The musical abilities of our closest relatives, gorillas and chimpanzees, are nil. Besides us, there are only four families of singing primates, and it seems that the ability has evolved independently in each case. These are the howler monkeys, the marmosets, the tarsiers, and the indris.

Of these, the howler monkeys are our closest relatives. Their song has a territorial function, as it warns other howler monkey herds about their presence in the forest, in order to avoid confrontations. But the songs are used mostly for the social life of howler monkeys, who sing several times per day, every day, each song lasting about 15 min. Often, a couple sings together; other times, the whole herd sings together, using polyphony and harmonies worthy of Bach himself. The last phrase of the song always ends in a crescendo, performed by a female (who has the last word). It seems that in this case the song has mainly to do with social cohesion.

To find so many examples of musicality in so many animals on our planet is encouraging, and it suggests that, maybe, music is used around the universe. On our planet, animal music mostly fulfills the double function of courtship and territory defense (using usually different songs for each function). Clearly this is the meaning of their music. Thus one could expect that, if our extra-terrestrial neighbors do use music, they will probably use it with this double function of territoriality and reproduction, too.

But what about human beings? Human music has some things in common with animal music. We also use music for courtship; and anthems (of nations, football teams...) are clearly territorial. But there are also important differences. The music of animals is always a long distance communication, while human music does not seem to be especially related to distance. Another important difference is the rhythm, the feeling of time passing, which is one of the most notable characteristics of our music (allowing us to dance to the music). In comparison, rhythm in animal music is usually irregular or absent. Finally, the functions of courtship and territoriality do not seem to be the most important functions in our music. In fact, only a small fraction of the music that we hear (or that is composed) has these functions.

This is not the meaning of human music. Its real purpose seems to be social – concretely, to communicate and transmit feelings. This is the best thing our music does. Listening to music makes us feel pleasure or perhaps sadness; it makes us happier, it frightens us, it makes us nervous. It is a way to communicate emotions. As Leonard Bernstein said, "Music is emotion," but human emotion.

Therefore, would E.T. appreciate Beethoven's music? Probably not.

Music and Language

We have seen that music has nothing to do with abstract communication. Music is not designed to exchange information, and some unnatural tricks are needed in order to achieve this purpose. Music is good to communicate emotions (music is very efficient in this task), of course, but not information.

Despite this, it is possible that music has a lot to do with human language. Maybe it is related to its *origin*.

In fact, the idea that the origin of human language lies in our ability to sing goes way back and can be traced to Jean-Jacques

Rousseau. All of the biological equipment needed for singing is the same one that is needed for talking. Today there are several researchers, such as Mario Vaneechoutte and John R. Skoyles, who think that human language had its origins precisely in the ability to sing and not the contrary. Singing came first. You do not need a vocabulary to sing; you can sing without words (maybe the ship-wrecked Alexander Selkirk sang on his island despite his forget-ting how to speak). To pass from song to speech you have to add some things – syntax, vocabulary, and so on. So it seems logical that song came before the speech, since it is simpler.

In fact, our bodies seem to have been "designed" to sing. Our organism has unique adaptations very similar to those of songbirds. Humans breathe in a way that is unique among primates, in that we are able to modulate very accurately the movements of the breast-bone. For example, chimpanzees can vocalize sounds, as "hoo hoo hoo hoo," but they must breathe between each vocalization. One breath, one vocalization. We do not need to do this. This kind of breath control is found only in songbirds (even howler monkeys can-not do what we can do). The anatomical characteristics of our vocal tract are more intimately linked to our ability to sing than our abil-ity to speak. For example, to sing we have to use our vocal tract completely, but we can talk without using parts of the vocal tract. Among all the sounds human beings can generate (and more than 700 different phonemes have been identified among all the languages of the world), the speaker of a given language uses only a small subset of them, underutilizing the capabilities of the vocal tract. But when the person sings, the range of sounds used is much wider.

Following this theory, we should be descendants of singing apes that, similarly to howler monkeys, developed song as a way to social-ize. By means of song they were able to communicate to the group their emotions – how they felt – without needing words (besides, it would also fulfill the traditional functions of courtship and territo-riality). This matches rather well with theories George Lakoff and the cognitivists defended: that language is produced, originated, or even made possible in first place by psychological aspects mainly related to basic emotions, that emotion comes before speech.

A possible proof of the fact that, before we talked, we were sing-ing apes, is how parents talk to their young children. We humans have a way of speaking that we use *exclusively* when talking to babies – it seems that we sing. And that, according to the theory, is precisely what we are doing. That kind of speech would be a living fossil of this phase in our evolution.

One of the virtues of this theory is that it could solve the paradox of human language that we encountered before, the contradiction about language being at the same time innate and acquired. In fact, what would be innate is our ability to sing. Therefore, babbling babies would not be innately trying to talk, as it seems, but learning to control their vocal tract for singing. In fact, the first vocal manifestations of children are more related to rhythm and intonation than to exclusive elements of language such as vocabulary or syntax. In pre-school classes, including classes in music education, it is normal to see 2-year-old children who still do not talk be able to hum melodies with no problems.

Our remote ancestors would have used music to communicate their mood. From the communication of feelings to the communication of concepts maybe it was just a single step. If those singing ancestors also had in their brains the protolanguage theorized by Bickerton (the one that, in theory, we share with the big apes, and that the common ancestor to them and us should also have had), to make this step could be rather easy[1]. But because they did not come from a tradition of talking beings, they had to invent words. Something acquired. But how do you invent words from scratch?

Ferdinand de Saussure defended the principle of the linguistic sign arbitrariness. In his own words: "The connection between the signifier and the signified is arbitrary; given that we understand by sign the total resulting from the association of a signifier with a signified, we can simply say: the linguistic sign is arbitrary. Thus, the idea of South is not linked by any inner relation with the sequence of sounds s-a-U-θ which is its significant; it could be perfectly represented by any other sequence of sounds."

---

[1]Being true to this theory, from fossil remains the most we could deduce is that Neanderthals sang, not that they talked. Although the complexity of their technology seems to imply the spoken interchange of abstract information, we should not dismiss the possibility of cultural transmission by means of imitation.

But today we know that this arbitrariness is not real. The mechanism to name objects is not completely arbitrary. Those first inventors of words had some guides to follow. For example, onomatopoeic words are abundant in all the languages of the world. This kind of word names objects by imitating how the object sounds (for example cuckoo, kiss, ratchet, to ring, to click, to flush...). It is an obvious way for naming things. In the origin of language there should be a big fraction of onomatopoeic lexis.

On the other hand, in a famous psychological experiment by Vilayanur S. Ramachandran and Edward Hubbard (although originally designed by Wolfgang Köhler), several subjects were asked to identify which of the following images was named booba and which kiki.

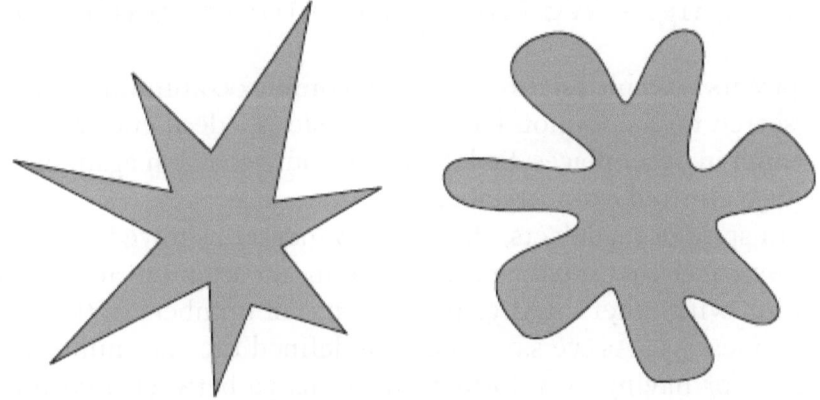

From 95 to 98% of the subjects of the experiment picked the jagged orange shape as kiki and the curvy violet shape as booba. Maybe this is because the lips take on a curvy shape when they say the name of "booba," or maybe because K has a sound harder than B (note also that the shape of the letter K is more similar to the first drawing and B to the second)[2]. In any case, what this

---

[2] On the other hand, in tests done by this author, presenting the same images but using changed versions of the names (bibi and kooka), most people chose bibi for the orange star and kooka for the violet cloud. Maybe the choice had as much to do with the vowels that are used as with the consonants. On the other hand, the results of Köhler's experiment are not reproduced in individuals with autism. The percentage of assignation in this case is close to 50%, that is, as assigning names randomly.

experiment shows is that the naming of objects is not completely arbitrary. Somehow, the human brain is able to establish links between shapes and sounds. For the first speakers, not only the sounds of objects, but also their shapes, could have been a guide for naming objects.

To finish this discussion about the relationship between speech and music, let us interject this thought, without any scientific foundation. Dolphins talk (or at least they do something very similar to human speech). On the other hand, their close relatives, the whales, sing. Did dolphins start as singing animals, too, making the next step to talk? Is music one of the ways towards complex communication?

What Might We Have in Common with E.T.?

Despite its possible relationship with complex communication, in our search for a common language, music is a dead end. We have to search in other places. So let us focus our attention again on the message emitted from Arecibo.

In spite of its defects, the Arecibo message showed us one of the keystones that must be considered in our attempts at communication with other intelligences: the use of numbers as the basis of the message. As we saw when we defined the communication process for having an information exchange between two intelligences, it is necessary that both parties share a common code, a common language. But is it possible to develop a language to communicate with an extraterrestrial civilization that is completely alien in absolutely every possible aspect? Some scientists think so, if the development of such a language starts from a common basis. However, what can we have in common with them?

Of course, both groups live in the same universe, and we are governed by the same laws of nature. If these laws are included somehow in a communication language, this increases the probability that it is usable. (As we saw earlier, this consideration already appeared in the Arecibo message, although in a naive way.) The other thing in common is mathematics (and logic), thanks to its platonic quality of universal knowledge. The ratio between the perimeter and the diameter of a circle is always the same number (pi), whatever the size of that circle. Further, 41 is a prime number, independently of cultures, ideologies, or civilizations. It is precisely through the use of logic and mathematics that a structure and grammar for a common language can be developed.

Here is where the first "but" arises. If it is obvious that all living beings in the universe are governed by the same laws of nature, it is not so clear that extraterrestrial civilizations will share with us the same mathematics, not just the kind of mathematics that humans do (geometry, calculus, algebra, etc.) but maybe even the mere concept of mathematics itself. Maybe we are the only intelligent species in the universe that developed this. Maybe the famous universality of mathematics is a myth, after all.

Again, the only guide we have to assessing the possibility of intelligent extraterrestrial having any kind of mathematics is to turn our eyes to the animal life on our world. Of course, we will not look at our planetary companions for the possession of any kind of mathematics in the strict sense (excepting the dolphins, maybe?), but we can evaluate how good they are in the fundamental concepts that are the basis of what we use to build up mathematics – i.e., space, sets, numbers, distance, etc.

Mathematician Animals

The prospects are encouraging. We can begin with the fact that many animals, when they are put in front of two different bowls of food, realize which contains more, and of course, they try to eat it. You the reader may think this is an obvious thing, but you have to take into account that the notion of "bigger than" is a legitimate mathematical concept.

Moreover, in nature we find numerous animals (rodents, monkeys, birds, etc.) that exhibit the most elemental arithmetic ability – the ability to count. Hens know how many eggs are in their clutch. Laboratory rats can be trained to count external stimuli (such as whistle blows or light flashes), touching a knob as a response only after a given number of stimuli.

As a general result of these experiments, counting animals are able to discriminate among small quantities with high accuracy, although they lose this ability when they are required to choose between similarly large quantities. However, if those quantities are quite different, animals do manage to discriminate. In tests with pigeons trained to peck a given number of times in return for a reward, they consistently differentiate four pecks from five pecks, but they fail when differentiating 49 from 50; nevertheless, they distinguish between 40 and 50 without difficulty.

There are also animals that can perform elementary operations with integer numbers. The chimpanzee Sheba was trained to properly identify integer quantities with the corresponding Arabic numeral characters! That is, she perfectly understood that, for example, the written symbol "4" represents the quantity of four objects, whatever these were. In an experiment she was led into a room where there were oranges stored in two different places. Sheba had to go to both places, see how many oranges there were in each place, and afterwards indicate the numeral corresponding to the sum of both quantities. She accomplished this task perfectly on all occasions, proving that besides counting, she has the ability of adding. But the abilities of Sheba go further. It has been proven that she can add numerical characters up directly, that is, when she sees two written numbers ("3" and "2," for instance), she consistently signs the numeral corresponding to their sum (in this example, the "5"). It should be noted that chimpanzees are the closest human relatives, and these striking results have been obtained after an intensive training. What will we find in untrained distant relatives? (Fig. 10.1).

To study the abilities for performing mathematical operations on animals in the wild, we can make use of their confusion over wrong mathematical operations. In an interesting experiment by researchers Hauser and Carey, a group of wild macaques were shown two aubergines (objects that were new and interesting for them), which afterwards were put on a platform and hidden behind a screen. After this, the screen was removed, and in some cases there were the two aubergines, but in others there was only one aubergine (the other was removed from behind the screen). On those occasions when there was a wrong number of aubergines on the platform, the monkeys showed surprise and looked with puzzlement at the platform much longer than in the case when the number of aubergines was correct. That is, they realized that $1 + 1$ had to equal 2.

Fig. 10.1 Sheba the chimpanzee points to the number 5 after counting a like number of apples during a demonstration at Ohio State University's animal lab. Courtesy of Ohio State University

The experiment was repeated for subtraction. After showing two aubergines and putting them behind the screen, one was removed from behind the screen in an obvious manner. On those occasions, when after removing the screen, there were two aubergines on the platform, the observing time for the macaques was again much longer than in the cases when there was the correct number of aubergines (i.e., one). This means that the wild macaques also realized that the correct result of the operation 2 – 1 has to be 1. This kind of experiment has been repeated with identical results with tamarins, some small new world monkeys related to titis, which are genetically much more distant from us (Fig. 10.2).

Fig. 10.2   Macaques group during mutual cleaning. Macaques are close relatives of ours with surprising arithmetic abilities

Mathematics and Darwin

Summing up, many animals in laboratory tests and during their natural behavior in the wild show the most basic arithmetic abilities (to count, to add, and to subtract). This evolutionary convergence among so many diverse species indicates that there is a selective pressure in favor of the acquisition of such elemental mathematical capabilities. In fact, to be able to count, add, and subtract offers a clear advantage to the species that can do it. We can appreciate this with a simple example.

Let us imagine this scene: an animal coming back to its warren sees four wolves entering it. Confronted with this menace, the animal remains outside, waiting for the predators to leave. As the wolves are exiting the warren, if the animal has mathematical abilities, its brain could reason as follows: "OK, four went in and now one has come out; thus three still remain. Hey, another has come out, so there are two still inside. Another one is coming out, so only one remains! Well, there goes another one, the last one! None remain inside, and thus I can go safely into my warren." If the animal does not have this mathematical ability, the mental process will be more or less something like: "Well, a pack of wolves went into my warren. There goes one! Let's wait a little more. There goes another one! Good, now there are less. Another one goes! Well, I think all of them have left so I'm going to go into my warren." Oops! As we see, the animal with abilities for counting, adding, and subtracting are naturally selected for survival.

Speaking about other basic physical-mathematical concepts such as space, distance, or the notion of "higher or lower than," any animal with mental abilities analogous to these will have an obvious advantage compared with those that do not, for example, when a predator has to anticipate the movement of its prey (or vice versa), or when choosing the fruit tree with a bigger food supply, or when calculating distances in a risky jump. The existence of such selective pressure, and the consequent evolutionary convergence that is produced in different animals is a strong support to the possibility that similar mathematical concepts can exist in the minds of extraterrestrial intelligent beings, supporting the use of mathematics in the development of a communication language.

However, not all mathematical concepts provide selective advantages. Therefore, the probability of sharing other mathematical notions with other animals or with extraterrestrial intelligences is lower. For example, humans can count, add, and subtract small quantities almost innately, but we are forced to memorize the multiplication tables, because there was no selective pressure favoring this ability. Thus, the ability for multiplying numbers has to be constructed subsequently from other mathematical basics that we do have.

It is relevant to relate here an interesting study using genetic algorithms. A genetic algorithm is a computer program that helps optimize work through a "Darwinian selection" process, analogous to what we find in the wild. The program is faced with a problem, and it gives a solution, a response. Initially, the program produces child programs, copies of itself but with slight modifications, randomly included here or there, in order to simulate the mutation processes undergone by living beings' DNA. These child programs are faced again with the problem and produce their own response. Those that give an answer distant to the expected one are deleted.

The remaining programs will be the parents of the next generation. Each new program generation will be closer and closer to the ideal result, as a consequence of this Darwinian selection process.

The case we are interested in is a computer program written by John R. Koza from Stanford University, who was faced with real planetary coordinate data taken on different occasions. The expected response was a mathematical formula that could predict the planetary positions. Of course, the first generated formulae were nonsense. The next generation produced formulae that were slight random modifications of their parent programs. For example an x^2 in the parent program could became an x^3 in one of the child programs, or an $(x+1)^2$ in another. Among all the child programs, only those that better predicted the orbital positions were allowed to survive, becoming the parents of the next generation. Within only 50 generations a computer program appeared that was exactly Kepler's Third Law, written according to Newton's formalism.

What can we conclude from this? That if our species had had selective pressure for predicting planetary positions, a matter of life and death if you will, surely nowadays we would have innately implemented in our brains Kepler's laws, just as we have the ability of adding and subtracting.

Summing up, maybe alien mathematics will share with us integer number arithmetic but little more. This conclusion is reminiscent of what finitist mathematicians claim, that the only valid mathematical knowledge is the one that can be deduced from integer numbers. The German mathematician Leopold Kronecker, head of this movement, even said that "God created the natural numbers, all else is the work of man."

A Cosmic Language

Our exploration of the mathematical abilities of animals seems to hint that a civilization able to develop radio telescopes has to have some kind of mathematics. Maybe different from ours, but mathematics nonetheless. Therefore it is a good option to base an interstellar communication language on mathematics. In fact, this may be the only option.

Using this approach, several attempts have been made, the most promising being Lincos, an acronym for the Latin expression *lingua cosmica*, an interstellar communication language created in 1960

Fig. 10.3   The mathematician Hans Freudenthal (1905–1990), a real E.T. talker

by the German mathematician Hans Freudenthal and published in his book *Lincos: Design of a Language for Cosmic Intercourse, Part 1*. In fact Lincos is an expansion of a former "cosmic language" called astraglossa, invented in 1953 by the British mathematician Lancelot Hogben. Astraglossa was a formal language designed for teaching mathematics to the hypothetical aliens. Lincos, on the other hand, is a richer and more powerful language that can be used to communicate complex non-mathematical concepts (Fig. 10.3).

The phonemes of Lincos are radio signals (or radioglyphs, following Hogben's nomenclature): different "whistles" with different meanings, which have to be deduced by the receiver. Lincos's structure is designed in such a way that it is a language that teaches itself, and given that actions speak louder than words, what could be better than learning to speak a little of Lincos ourselves? Thus, let us dust off our radio telescope, tune it to the proper frequency, and connect it to a radio with a speaker to listen to the lessons of the course "Learn Lincos in Minutes."

The first lesson that will arrive at the receiving end will be the following:

. .. .

Where the symbol · in fact represents an elemental radio "bip," and the white spaces, pauses. Time goes from left to right, in the reading direction. That is, in our radio, that signal would sound something like: bip, bipbip, bipbipbip, bipbipbipbip... With a little effort, we could deduce that the contents of this first lesson are the natural numbers from 1 to 9. That is, that "bip" signals represent natural numbers, which in some sense seems very much like the signal Nicola Tesla thought he received from Mars.

Once we understand this, the second Lincos lesson arrives. It sounds like this:

$$\cdot \; \mathcal{H} \cdot \; \clubsuit \; \cdot\cdot \qquad \cdot\cdot \; \mathcal{H} \; \cdots \; \clubsuit \; \cdots\cdot\cdot \qquad \cdots \; \mathcal{H} \; \cdots\cdot\cdot \; \clubsuit \; \cdots\cdots\cdot$$

On this occasion what are represented with $\mathcal{H}$ and $\clubsuit$ are two new kinds of radio whistles, which maybe in our receiver sound like "mook" and "crack," respectively. That is, the first sentence of this second Lincos lesson could be something like "pip mook pip crack pip pip."

To assure the understanding of these new radio signals, we would have to send several examples, as we have done here. This former lesson is a little harder, but it is worthwhile for the reader to try to deduce the meaning before proceeding.

Did you succeed? Have you seen the relationship? If your intuition is working, you would have deduced that the $\mathcal{H}$ radioglyph represents addition, that is, the sign +, and the $\clubsuit$ radioglyph means "equal to," that is, the sign =. Therefore we have learned two new Lincos words. Try now with the following lesson to deduce the meaning of two new radioglyphs, shown below. Take your time:

$$\cdot\cdot \; \clubsuit \; \cdot\cdot \; \bullet$$

$$\cdot \; \mathcal{H} \cdot \; \clubsuit \; \cdot\cdot \; \bullet$$

$$\cdot \; \mathcal{H} \cdot \; \clubsuit \; \cdots \; \blacksquare$$

$$\cdot\cdot \; \mathcal{H} \; \cdots \; \clubsuit \; \cdots\cdot\cdot \; \bullet$$

$$\cdot\cdot \; \mathcal{H} \; \cdots \; \clubsuit \; \cdots\cdot \; \blacksquare$$

$$\cdots \; \mathcal{H} \; \cdots \; \clubsuit \; \cdots\cdots\cdot \; \bullet$$

$$\cdots \; \mathcal{H} \; \cdots \; \clubsuit \; \cdots \; \blacksquare$$

$$\cdots\cdots\cdot \; \mathcal{H} \; \cdot\cdot \; \clubsuit \; \cdots\cdots\cdot\cdot \; \bullet$$

$$\cdots \; \mathcal{H} \; \cdots \; \clubsuit \; \cdots \; \blacksquare$$

OK, starting from what we learned in the previous lesson and using simple logic, we can easily conclude that these two new Lincos "words," ● and ■, represent respectively the concepts "true" and "false" (or "correct" and "wrong"). Once we have these two concepts, the conversation can be enriched, and it is easier to include new concepts, as in the following example:

$$\cdot \; \S \; \cdot\cdot \; \bullet$$

$$\cdot \; \S \; \cdots \; \bullet$$

$$\cdot\cdot \; \S \; \cdots \; \bullet$$

$$\cdots \; \S \; \cdots \; \blacksquare$$

$$\cdots \; \S \; \cdots\cdots \; \bullet$$

$$\cdots\cdots \; \S \; \cdots\cdots \; \bullet$$

$$\cdots\cdot \; \S \; \cdots \; \blacksquare$$

In this case, taking into account the concepts of true and false, with a little effort one can deduce that the new radioglyph § represents the concept of "lesser than." We will finish the course with a last lesson, in this case purely in the field of logic, introducing two new Lincos words:

$$\bullet \; \blacktriangle \; \bullet \; ❖ \; \bullet$$

$$\bullet \; \blacktriangle \; \blacksquare \; ❖ \; \blacksquare$$

$$\blacksquare \; \blacktriangle \; \blacksquare \; ❖ \; \blacksquare$$

$$\blacksquare \; \blacktriangle \; \bullet \; ❖ \; \blacksquare$$

$$\bullet \; \blacktriangledown \; \bullet \; ❖ \; \bullet$$

$$\bullet \; \blacktriangledown \; \blacksquare \; ❖ \; \bullet$$

$$\blacksquare \; \blacktriangledown \; \blacksquare \; ❖ \; \blacksquare$$

$$\blacksquare \; \blacktriangledown \; \bullet \; ❖ \; \bullet$$

The solution, of course, is that ▲ and ▼ respectively represent the conjunctions "and" and "or." This is the scheme proposed by Freudenthal in his book. With consecutive lessons new definitions are introduced, until by the end of the book the vocabulary is so complete that communication of not only mathematical concepts but also any kind of information is possible.

Nevertheless, the process is slow. Since Freudenthal makes few assumptions to start, it is necessary to communicate a great amount of information about the language and its working before being able to communicate interesting information. To speed up the process, the mathematician Carl DeVito, together with the linguist Richard Oehrle, thought about incorporating more fundamental science into the language in order to give a context where mathematical definitions could be inserted into Lincos. They presented their ideas in their 1990 paper, "A Language Based on the Fundamental Facts of Science." They assume certain fundamental scientific facts to be known by any radio telescope-building civilization: being able to count, understanding chemical elements, knowing the melting and boiling points of different pure substances, and knowing the properties of the gaseous state. Using this supposedly common knowledge, it is possible to speed up the language learning process and communicate interesting information earlier. From this, they developed their own communication language, a modification of Lincos. Given that DeVito and Oehrle did not name their language, we will take some liberties here and call it Lincos 2.0.

Other interesting works in interstellar communication appeared as a result of Lincos. Here are the two most interesting. The first is certainly a novel concept, born from the computational capacity of computers. The idea is to send an algorithm (a program) so that when the receivers run it, it will teach them about our world. This would allow a level of interaction impossible if communication were based only on the exchange of passive messages between them and us (given the distances and times that separate us). This is the case of CosmicOS by M.I.T. engineer Paul Fitzpatrick. Of course, this imaginative approach assumes, on the other hand, that the receiver civilization has to have computational capacity. To understand the message, they would have had to develop something equivalent to our computers, something that we have had for only eight decades.

The second idea is the development of a visual "Lincos," à la Arecibo, that is, to send images by radio in a similar manner to the Arecibo message, the content of which consists of a kind of self deducible language such as Lincos (only with written characters instead of radioglyphs) and including images. This is precisely the option chosen in the Carl Sagan novel (and subsequent movie) *Contact*, where inside the radio message, a whole book was encrypted page after page, with symbols, images, diagrams, and other graphics (the first lessons had an amazing similarity with the ones of our Lincos course). Similar messages have been developed and were already sent to the stars. In 1999 from the deep space antenna at Evpatoria (Ukraine) a message, called *Cosmic Call*, was transmitted to four Sun-like stars. The message consisted of 23 radio sequences, each one made up of 16,129 pulses. The entire message was transmitted three times to each one of these stars, during a period of 3 h. In 2003 the same message (with slight modifications) was sent again to another five stars.

When seeing the curious number 16,129, the attentive reader immediately asks if it is associated with prime numbers, and the answer is yes. The number 16,129 is equal to 127×127; thus, each one of those 23 sequences is in fact a square image made up of 127×127 pixels, that is, a page of the message. When these pulses are rearranged in a way similar to the Arecibo message, we get pages like the following (Fig. 10.4):

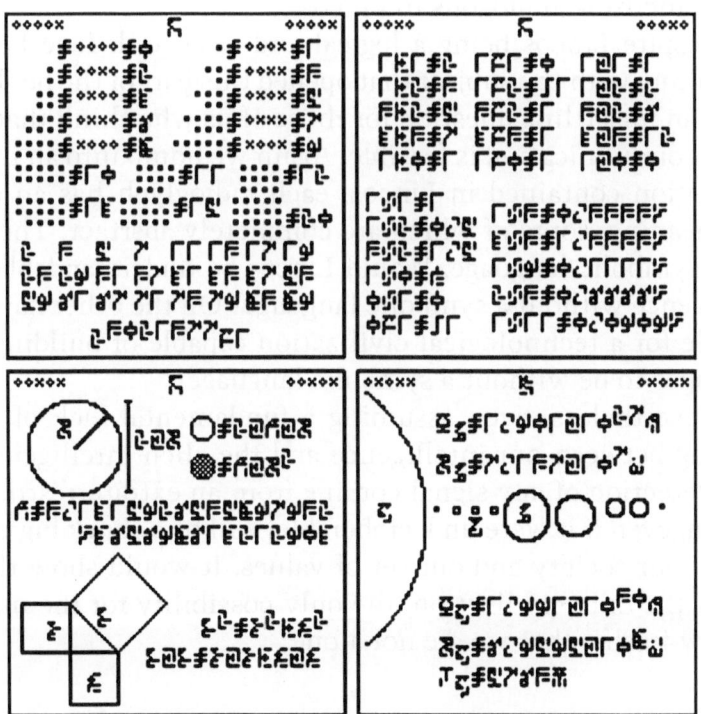

Fig. 10.4 Pages 1, 2, 5, and 11 of 1999 Cosmic Call. (Courtesy of Yvan Dutil and Stephane Dumas.)

Each one of these pages is surrounded by a black frame, which only appears if the page has been properly decoded. As we see, the first pages are basic lessons in arithmetic, and afterwards more elaborate concepts are introduced, such as a Pythagoras theorem or the shape of the Solar System. Of course, the drawbacks for this kind of information sending are the same ones as for the Arecibo message: there are different ways to rearrange 16,129 elements in a 127×127 array, and it assumes that the message receivers will have visual capabilities equivalent to our own.

On the other hand, Lincos does not presuppose any requisite as to how the receiver perceives the world or which of their senses are being used – only that they have to be able to detect radio waves with their own sensors, which would already be designed according to their senses. Besides, Lincos does not have any problem about how to arrange the signal data, as they will be already ordered by the timing of the arriving radioglyphs. This "universality" of Lincos when faced with all the other invented interstellar communication media, biases scientists to think that, if sometime any kind of message from our galactic neighbors is received, it will be in something analogous to Lincos.

Despite Lincos being a logical and easy to deduce language and linking with the subject that opened this part of the book there is still an unsettling question for those of us who think that interstellar communication is feasible. Again we bump into an implicit assumption contained in Lincos: each radioglyph has an associated meaning, some of which are completely abstract. Therefore, it is a symbolic language. Would Lincos be understandable to an intelligence without a symbolic language? On the other hand, is it possible for a technological civilization capable of building radio telescopes to be without a symbolic language?

Nevertheless, even assuming a fundamental lack of understanding between our intelligence and the alien intelligence, the mere detection of any signal coming from an extraterrestrial civilization, even if it were undecipherable, will be a shock big enough to rock our society and our set of values. It would show that we are not the only civilization, the only possibility for the universe to know itself – that we are not alone.

11. Is Anybody Out There?

Fermi's Paradox

Are others actually out there? If so, why don't we know anything about them yet? Although this may seem a trivial question, it is not an easy one to answer at all.

This question is called the Fermi paradox, also known as the Great Silence. Its most usual formulation is in the form of an easy question that the well-known Italian physicist Enrico Fermi asked in the summer of 1950 during one of his frequent visits to Los Alamos National Laboratory (New Mexico). During lunch at the café one day he was speaking with his table companions, Manhattan Project physicist Edward Teller, Herbert York, and Emil Konopinski, about extraterrestrial civilizations and interstellar voyages. Fermi was well known for his ability to do good numerical estimates from little data. So, he did several quick calculations during the lunch and he posed this question to his colleagues: "Where is everybody?"

Why is this question a paradox? Mainly, because life naturally tends to expand; otherwise, it would disappear. Let us suppose that each generation of a species produces a number of descendants smaller than the number of progenitors. For example, let us assume each couple has an average of 1.5 children. The most elementary mathematics shows that for this species, each eight generations, the population will be reduced by 10%. Thus, if we start with an initial population of a million, at the end of 48 generations only one specimen will remain. That is to say, producing fewer descendants than progenitors, the species becomes extinct for a mere mathematical reason, independently of how well adapted to the environment the species might be.

F.J. Ballesteros, *E.T. Talk*; Astronomers' Universe,
DOI 10.1007/978-1-4419-6089-4_11,
© Springer Science+Business Media, LLC 2010

What if the species produces exactly the same number of descendants as the number of progenitors, that is, if each couple has exactly two cubs? Again, extinction is the destiny of the species, because having two cubs for each couple does not guarantee that those descendants will necessarily survive to the adult age and have in turn their own descendants. There are predators that will hunt some cubs and also some unavoidable accidents that will eliminate some of them before reaching adulthood. As time passes, in each generation fewer and fewer specimens will reach maturity, and finally the species will disappear.

Therefore, the strategy needed is to have more children than progenitors each generation. This is truly the strategy that all living beings on Earth follow. Those that for any reason have not followed it have disappeared. We ourselves have expanded across the whole planet, and the human population is becoming more and more numerous. Therefore, it is completely justified to suppose that the same strategy of overproduction of descendants is being followed by all living beings across the universe, wherever they may be.

So what? Well, if at a given moment, an intelligent species masters interstellar travel in an effective way and crosses the gulf between solar systems, due to this innate tendency to expand, they will start (extremely slowly, without doubt) to colonize new worlds they discover. These new worlds, once consolidated, can in turn be the focus for new colonizing expeditions.

For example, let us suppose that from the mother planet two colonizer expeditions depart, and that in a given moment, each one of these two daughter colonies send another two new expeditions that found four new colonies, which once consolidated, will send in turn new expeditions, etc... The progression of inhabited stellar systems after each colonization wave will be: 1 3 7 15 31... At the end of only 36 colonization waves, the number of colonized systems will be more than a 100,000 million. That is to say, approximately equal to the number of stars in our galaxy. In this scenario we have considered that the colonizing expeditions only depart from the last colonized systems, and in a quantity of two expeditions for each system. If we let the more ancient systems send new colonists also, and if they send each time more than two expeditions, that same number would be reached long before, in a smaller number of colonization waves (Fig. 11.1).

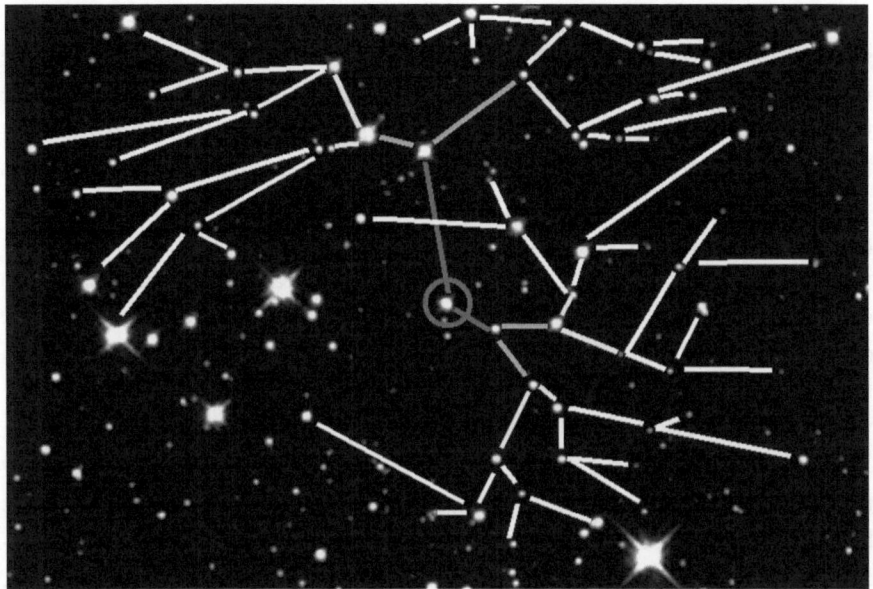

Fig. 11.1   Artistic representation of a process of galactic colonization for a civilization with interstellar travel, after five colonization waves. The solar system of the mother planet is *circled*

How long could it take for such a process of colonization to fill the whole galaxy? Let us make some calculations. From a technological point of view, it does not seem difficult for a spaceship to reach speeds close to 10% the speed of light, taking about 100 years to cross a 10-light-year distance. Thus, the closer stars (which are at a distance of about 5 light years) could be reached after about 50 years of travel. The relativistic expansion of time does not help to make the travel shorter. At those speeds, 50 years of travel will still mean 49 years and 9 months on board.

Once in the new stellar system, it will take a lot of time for the colony to settle and proliferate, as they would have to overcome numerous difficulties before feeling at home. Let us be generous and give each colony 5,000 years to consolidate, before being able to send new colonists to other systems. With these numbers, it is easy to calculate that each 25,000 years the colonies will cover a region of about 50 light years. As the diameter of our galaxy is about 100,000 light years, the colonizing civilization will take about 50 million years in covering the whole galaxy – an enormous amount of time. Almost a thousand times the history of our species.

But our galaxy is a place of enormous numbers. The age of our galaxy is about 13,600 million years, almost 300 times the time necessary to complete such a process of colonization. Although the first generations of stars had low metallicity, with planetary systems formed by simple gas balls, it is estimated that rocky planets similar to ours have existed for at least 9,000 million years. Earth, at only 4.5 million years, can almost be considered a newcomer. If we take into account how fast life appeared on Earth, it is reasonable to conclude that there must have been civilizations in the galaxy *thousands* of millions of years ago. The most ancient traveling civilizations have had time enough to reach us, truly, several times. Therefore, where is everybody?

This was what Enrico Fermi reasoned, more or less, during that meal at Los Alamos. Since then, solving this paradox has become a usual reference point among SETI scientists and astrobiologists, and numerous attempts have been carried out to solve it. All these solutions can be divided basically into two groups: those who allege that there are no other technological civilizations, and those who defend that there are but we still have not seen signs of their existence. Let us here do a quick review of the different attempts to solve the Fermi paradox.

Some Solutions

The first solution group includes the approach of "Rare Earth," which we already saw in the first part of the book. The mediocrity principle is false; our Earth is really an extreme rarity, and there is no life in anywhere else in the universe.

Alternatively, although unicellular life could perhaps be more or less common in the universe, our planet is a unique case, where only an enormous number of coincidences has led to the development of complex life forms. As we saw in the first chapter, on Earth the steps to multicellular life took a long time.

Another solution, leading to the same result, is the one that defends that perhaps life abounds, and maybe there is even multicellular, complex life on other worlds, but it has not developed intelligence. Taking everything into account, intelligence does not seem to be an imperative of evolution. For the last 670 million years of Earth's history, multicellular animals had gotten along quite comfortably without it (at least, at the level used by homo sapiens). Moreover, it seems indispensable to have the equivalent of a neural system develop. It does not seem possible that vegetable terrestrial life, in spite of being complex organisms, will develop any organ similar to a brain even in many millions of years. Therefore, according to both solutions, our world is the only corner in the universe where matter has become self-conscious and has begun to understand itself.

Similar, although with differences, is the solution that affirms that there is nothing special about Earth. It is just that, simply, we are the first ones. After all, some civilization had to be the first to appear. Although, if this is so, why has this happened so late, after 9,000 million years of existence of Earth-type planets? One explanation is that it took a very long time for multicellular life in the galaxy to appear. The enormous amount of time that, on our planet, passed from the appearance of life to the appearance of multicellular beings was not due to the fact that the transition, by itself, was difficult but to massive extinctions at a galactic level that impeded the transition, at least until about 1,000 million years ago. Are there truly galactic processes that can produce such extinctions?

The finger of science points to the main suspect: a strange phenomenon known as gamma ray bursts (or GRB in their abbreviated form). These are mysterious eruptions of gamma radiation with great intensity and very short duration (from a few minutes in the case of the short ones up to some hours), whose origin is still not clear. Its discovery put a tragicomic note to the Cold War. During the 1960s, the United States put a set of satellites (the Vela series) in orbit, in order to detect gamma rays produced by possible nuclear tests being carried out in the Soviet Union that violated signed agreements. Indeed, gamma ray signals were detected, but, to the military's surprise, they did not come from the Soviet Union but from space! After a terrible fright (they thought the Soviets were detonating nukes in space) scientists realized that it was in fact an astronomical phenomenon. In spite of this, the existence of GRBs constituted a military secret until 1973.

During the last few decades these bursts have been studied in order to reveal their origin. Nowadays we know they have a cosmic origin. Current telescopes have seen them in action inside completely normal galaxies, and recent data leads us to believe that they are some kind of hypernovae, incredibly intense supernovae (at least 100 times more powerful than standard supernovae) produced by the collapse of enormously massive stars. These gamma ray bursts are so violent that they can completely devastate the planets of the galaxy where they occur (to be more precise, the hemisphere of the planet exposed to the GRB), with the consequent ecological catastrophe. Therefore, only the simplest organisms would survive, impeding the step to multicellular life.

As time passes, the gamma ray burst's rhythm decreases, as the super-massive star that produces it disappears (this type of star was more common during the youth of the galaxy). If the time between two consecutive such explosions is big enough, it might permit complex life to arise (Fig. 11.2).

On the other hand, perhaps we are not the first intelligent beings of the galaxy but the first (or the only ones) that have developed a technological civilization. Maybe the true bottleneck

Fig. 11.2 Artistic representation of a gamma ray burst, or GRB, a kind of incredibly intense cosmic explosion, perhaps responsible for massive extinctions at a galactic level (Courtesy of NASA/SkyWorks Digital.)

is the development of technology, indeed. Perhaps, among all the intelligent beings in our galaxy, we are the only ones that, at the same time, have had instrument capabilities and developed a symbolic language. Or maybe, only our kind of mathematics can allow the technological development necessary to build space-ships and radio telescopes. Or perhaps the motive is even more fundamental: fire has been conquered only on our world, a previ-ous step that many scientists think indispensable for the start-up of a technological civilization, or maybe only on our planet did two phenomena occur at the same time – complex life conquered land, and the planet developed atmospheric oxygen.

Finally, among the solutions of the group "there are no other technological civilizations" is the one that Fermi himself pro-posed: as civilizations arrive to a certain technological level, they self destruct. Their own success is their doom. That is, there have been other technological civilizations, but they do not exist anymore. They disappeared before mastering interstellar flight. Nowadays only ours remains, and the clock is ticking.

Remember that Fermi raised the paradox in the middle of the Cold War and its arms race. He himself had taken part in the Manhattan project, which produced the first atomic bomb. The power of these artifacts, and the possibility of a nuclear war, justi-fied only too well this possibility. However, it is not necessary to have a nuclear war to get this scenario. Unfortunately, there are other causes that could produce the self destruction of a civiliza-tion, such as pandemics due to overpopulation, or the destruction and exhaustion of natural resources (also due to overpopulation). Moreover, technological development puts more and more power at the disposal of a single subject. Cases of young people creating computer viruses that cause important economic losses are becoming more frequent. Nowadays it is not so ridiculous that an individual can even possess an atomic bomb or an especially dangerous biological weapon. If this continues, a lunatic could end up having enough power to put an end to the whole civiliza-tion, as illustrated in the movie *Twelve Monkeys*. (Science fiction is an inexhaustible source of solutions to the Fermi paradox.) If the birth of civilizations is uncommon and their average lifespan in possessing sophisticated technology is short, the answer could be the silence that we observe.

But there are more optimistic solutions to the paradox. In the second group of solutions, "there are other civilizations but we do not know of them," the easier one says simply that technological civilizations in the galaxy are not colonizing. We are the exception. This is a difficult solution to believe, however, because, as we have seen, all life forms expand; it is a strategy for the survival of the species. Besides, this solution presents another kind of problem, which we can call the uniformity problem (and it is a problem that we will find in many solutions to the Fermi paradox). In this situation, absolutely all the civilizations in the galaxy must behave in the same way. No one can start a colonization process, because as soon as one starts, we would run right up against the Fermi paradox. But this type of uniform conduct is difficult to explain.

The more plausible solution is the one that bets on the impossibility of interstellar travel. The dangers implied in moving between stellar systems can be so high that they are successful only on rare occasions. Let us remember that if travel must be carried out in a reasonable time, the starship has to travel at enormous speeds, increasing the risk of collision with small interstellar objects. On the other hand, the economic cost for sending such an expedition is perhaps unaffordable for a civilization, and it makes more sense to send exploration probes, as was the case of the monoliths in *2001, A Space Odyssey* (although in that case, where are the probes?). Or perhaps once travelers reach a new stellar system they find it impossible to colonize, and the expedition fails.

If civilizations cannot extend out into their stellar systems, this would explain why the galaxy has not been colonized. It does not explain, however, the fact that we do not receive any emission from them. What happens to radio signals from these civilizations?

In the opinion of SETI scientists, if a massive colonization of the galaxy does not occur, inhabited worlds with intelligence will be scarce, making it easy for their signals to go unnoticed. It is only a question of time in finding their signals. At the present moment, not enough time has passed. More disturbing is the thought that perhaps their signals have already arrived but we did not understand them. Are they too odd to be recognized as such? Are in fact GRBs the results of the industry of incredibly advanced Type II or III civilizations? Moreover, we are only looking within the electromagnetic spectrum. Are they using perhaps gravity waves or neutrino fluxes, as in the novel *His Master's Voice* by Stanislav Lem? Although, on the other hand, why should they use a form of radiation that is difficult to manipulate, having at hand electromagnetic waves, which are much more manageable? Also, to hypothesize that no other civilization is using electromagnetic waves when they could, comes up against the uniformity problem, as all of them have to behave in the same way.

Or it is possible that everybody is listening and nobody is transmitting? After all, we are not making a systematic campaign of endless radio emission to space. We have just sent a few sporadic messages, like that from Arecibo or those two Cosmic Call campaigns, during short time lapses. The real possibility that some of these signals will ever be detected is rather tiny. Perhaps it is the same with them. But again, this falls into the uniformity problem. Besides we have to take into account that we are indeed endlessly sending signals into space, although in an involuntary way. Our radio and television signals flow outwards without interruption. Therefore, even if other civilizations are not making an active campaign of cosmic communication, we will detect sooner or later some of their internal use emissions.

Other solutions to the Fermi paradox include accepting the possibility that civilizations are indeed extending across the galaxy. The example that opened this chapter described a colonization process that grew exponentially, but there are people that defend the existence of mechanisms that, even while permitting a growth in the number of colonies impede exponential growth. For example, it is possible that a civilization with space travel will not pass to the next star until all the resources of the current stellar system are consumed, leaving it afterwards, in a similar way to the *Independence Day* movie. Thus, the growth of visited planets is linear, not exponential. However, this solution has two problems. One, the uniformity problem again: all civilizations with interstellar travel must behave the same. The other, they have to enact strict birth control laws to insure that the population is always within the limits a stellar system can maintain.

The physicist and science fiction writer Geoffrey Landis has thought up an ingenious solution in which the number of colonized planets grows, but in a fractal way. The key is that not all colonies become new sources of colonization; some of them lose the interest to keep colonizing. Depending on what the probability is that a colony becomes a source of new colonies, the evolution of the process will be different. If this is high, the growth will be exponential (as in the initial example), and if it is very low, the process of colonization will finally come to a stop. If it is exactly equal to a critical value, the number of colonies increases linearly, producing a fractal pattern, with big empty areas that will never be colonized (Earth being inside one of them). Unfortunately, this ingenious theory does not give an explanation to justify why the formerly mentioned probability must be equal to that critical value.

Both cases have the same weak point. Where are the emissions and artifacts of these civilizations? A linearly growing colonization process, even if it has not reached our world, will have some devices (such as ring worlds, Dyson spheres, optical or radio signals, or tracks of antimatter engines) that should be observable with current telescopes. Up until now, nothing like this has been observed.

All the solutions we have seen until now assume the absence of interstellar visitors on Earth. What if this is not true? Clearly, the trivial solution to the Fermi paradox is to conclude that, indeed, we have been visited by star travelers. But is there any proof of this?

Surely the most commented case regarding possible alien visits in the past is the one concerning the African tribe of Dogons. In 1931, the French anthropologist Marcel Griaule visited this tribe and remained fascinated by their unique customs, so he made several visits to study the Dogons over a period of several years. From his investigations of these people, he published an article in 1965, "Le Renard Pâle," in which he said the Dogons had a disproportionate amount of astronomical knowledge, taking into account the poor technical media they had. According to Griaule, the Dogons knew that Jupiter had four main moons, that Saturn was surrounded by a ring, and that the Moon was a dead and dry world. More disconcerting still was the fact that the Dogons affirmed that their own culture came from the star Sirius (according to Griaule, the center of Dogon religious life), which they called Sigu Tolo. They also affirmed this star was being orbited by a very tiny star "composed of the heaviest metal of the universe," which they

Fig. 11.3 The Dogon country, in Mali. Not so backward as they say

called Po Tolo. A third star, called Emme Ya, would be rotating around the whole system (Fig. 11.3).

What has astronomy to say about this? Well, indeed, Sirius is being orbited by a star (named by astronomers with the poetic name of Sirius B), which happens to be a white dwarf, an object with very high density. This coincides rather well with the description of Po Tolo. Moreover, in 1995, astronomers of the Nice Observatory announced they had detected traces of the existence of a third star in the system of Sirius (maybe Emme Ya?).

All this astronomical knowledge should have remained out of the Dogon's reach, as it requires the use of astronomical instruments they did not have (in the case of Sirius B, very powerful telescopes are needed to see it). How did the Dogons acquire all this knowledge? The answer could be that it was provided by an expedition of aliens coming from Sirius, who visited Earth in a remote past. This is the answer usually chosen by those who popularized "the Dogon mystery." Unfortunately for the mystery lovers, this is not the only possible answer.

Firstly, Sirius B had been already discovered in 1844, and its status as a white dwarf was determined in 1915. In fact, it was the first white dwarf discovered. This was a true astronomical bomb-shell, and the dense companion of Sirius was front page news in numerous publications at the beginning of twentieth century. Even the suspicions that the system could have a third star are old, dating from 1894, as the system presents certain orbital irregularities. That is to say: all the astronomical knowledge the Dogons reported to Griaule were part of the current astronomical knowledge corpus at that time.

Secondly, this supposed Dogon knowledge was not free of errors. The Dogons (according to Griaule) identified Saturn as the most distant planet from the Sun, ignoring the existence of Uranus and Neptune. If the knowledge of the Dogons really came from interstellar travelers that arrived in our Solar System, it seems unreasonable that these giant planets went unnoticed by them. And regarding the possible discovery of Sirius C in 1995, it is necessary to emphasize that the astronomers of the Nice Observatory say they had detected *signs* of the presence of a star, not the star itself (in fact, their article is entitled "Is Sirius a triple system?"). Moreover, according to this scientific study, in order to justify the orbital irregularities, the third star has to orbit exclusively around Sirius, in an orbit very close to the star, and not around the Sirius–Sirius B pair (as it is supposed that Emme Ya does). Subsequent observations carried out by the Hubble Space Telescope of the Sirius system, following the conclusions of this paper, have so far not found any trace of the existence of that third star.

Thirdly, the Dogons are usually portrayed as an isolated and lost tribe in the innermost regions of the African continent. This is not true. In fact, they are very much in communication with others. The Arab expansion expelled them from their original lands, and during the following centuries they have lived together with their Muslim neighbors (in fact, some of them are Muslim, and others are Christian). They were recruited as soldiers for the colonial forces. And at the beginning of the twentieth century, there was missionary activity and numerous French schools in the area, which could have certainly put them in contact with the astronomical knowledge of that time.

Moreover, Griaule himself was an amateur astronomer, and part of the problem of the Dogon mystery seems to lie in him. This French anthropologist did not speak the language of the Dogons, and all his field work was done through translators and intermediaries, usually belonging to the French colonial army. The Griaule methodology consisted of bringing members of the Dogon people to his camp and, once there, through an interpreter, ask them a series of questions. By using this methodology, it is easy for information to suffer modifications and (involuntary) distortions according to the previous knowledge and expectations that Griaule might have had, thus leading to a reconstruction of the information by the French anthropologist.

In fact, subsequent research carried out in Dogon land have never found even a vestige among these people of such detailed astronomical knowledge nor an understanding of the relevant position of Sirius. No anthropologists following Griaule's trail have reproduced his surprising results. Moreover, the Belgian anthropologist van Beek who, from 1979, spent 11 years with the Dogons searching for evidence of Griaule's affirmations, found that the Dogons had not even heard of Sigu Tolo nor knew that Sirius (which they called Dana Tolo, and not Sigu Tolo, as Griaule said) were a double system. Thus, the extraterrestrial origin of the "Dogon mystery" was ruled out.

Carl Sagan, in his book *Intelligent life in the Universe*, proposed another case that has many of the conditions we would hope to find if, in historical times, there had been true contact with an extraterrestrial civilization. It is the Sumerian legend of Oannes. This legend, dating from approximately the year 4,000 BC, has come to us through intermediaries, since no Sumerian original narrating the legend has survived to this day (although there are abundant representations of Oannes). The only sources are the writings of the Babylonian priest and writer Beroso, a contemporary of Alexander the Great and, thus, writing about 3,500 years after the genesis of the Oannes legend. Beroso was in contact with the original Sumerian cuneiform texts that contained the legend, and he used them to write his "Babylon history." Unfortunately, Beroso's book has not survived either, and we only know of its existence through the writings of other historians of antiquity – Alexander Polyhistor, Abydenus, and Apollodorus of Athens – that referred to Beroso's work and its content. Through their writings it is possible to reconstruct rather well Beroso's original text.

In the text it is said that, after the creation of the world, men lived in Mesopotamia "as beasts of the wild," until a day when, from the Eritrean Sea in the Persian Gulf, appeared an *anedot*, an "animal endowed with reason" that could speak, called Oannes. Oannes was devoted to teaching the Mesopotamian inhabitants all knowledge: "the literature, the sciences and all kinds of arts. He taught them how to build houses and temples, how to compile laws and the geometry, how to plant seeds and gather their fruits. [...] His teachings were so universal that, since then, nothing has been added to improve them."

Oannes was easily identifiable in the Sumerian engravings; he was depicted "as a fish, but under his fish head he had a second human head, and joined to his fish tail he had human feet," and he was described as having amphibian habits: "He got out of the sea at sunrise to teach men and to speak with them, although he did not take any food [...] and he again dived in at sunset, remaining the whole night in its depth."

As generations passed, other anedots appeared from the sea, five altogether, all of them with the same curious aspect, all of them knowing the work of their predecessors, and all of them with the same task: to civilize Mesopotamia (Fig. 11.4).

It is almost certain that Oannes and his people represent sea divinities, in the style of Neptune and the Tritons, although it is curious that they are defined as "animals" (in other excerpts, as "half-demons," and even as "things") but not as gods. It is difficult to escape the charm of this legend, and one can easily be seduced by the idea that the first civilization that appeared in our world, between the Euphrates and the Tigris, had a stimulus from alien visitors established in a submarine base under the Persian Gulf, that took us to our current civilization – an idea very much in line with the short story "Encounters in the Dawn" by Arthur C. Clarke.

Finally, even if neither of the two previous cases were real narrations of an encounter with interstellar travelers, that does not mean that they were not here. Perhaps their visit was very

Fig. 11.4   Printing of a mesopotamian cylindrical stamp where Oannes appears represented (right)

early, before humans evolved. Perhaps someday archaeology will provide us with a surprise, finding an impossible object in a very ancient stratum. Some astrobiologists even speculate that maybe our world is already one of their colonies. That is, that aliens came, and we are them. We know without any doubt that the whole gamut of animal and vegetable life on our planet appeared and evolved on Earth, but maybe the first organism, the primordial cell that started everything, was sown on our planet by these visitors, in a kind of voluntary panspermia. This would be a way to colonize worlds with little risks for the colonizer (better said, "sower") civilization.

Really, though, this is not a solution to the Fermi paradox, given that, as we have already seen, there has been plenty of time for our planet to have been visited on several occasions after the first visit of the seeding expedition. Where are the traces of those other visits? Besides, all the proofs that terrestrial life has evolved completely on our planet from prebiotic chemical processes (i.e., biochemical relics of the RNA world) play down the strength of the hypothesis that life was planted on Earth fully formed.

For the sake of completeness, we will conclude by giving a quick review of other solutions that are completely outside of the field of science. For example, a well known solution is the one affirming that aliens are already here but they are hidden, because of an ethic of non-interference with emergent civilizations, in the vein of *Star Trek's* First Directive, or because for them we are a kind of "terrestrial zoo" to study, or (as in *Men in Black*) because the government is hiding them.

This is a rather paranoid solution that links very well to another, which affirms that interstellar visitors are here...and that they are the UFOs. However, after several decades of study of the UFO phenomenon, the truth is that there is little objective proof linking this "phenomenon" with space travelers, but rather with psychology, or even psychiatry. Truly, stories about UFOs, landings, and abductions suspiciously appear to be extremely similar to the ancient stories of fairies and forest elves (or to Marian appearances), informing more about how the human psyche works than about life in the universe. The anti-scientific, paranoid behavior, completely rigor-lacking, tending to self-deceit, and, in many cases opportunistic and even contemptible, of the UFO phenomenon defenders and researchers, makes "ufology" fall completely in the classification of pseudoscience.

To finish, a disturbing and unpleasant solution to the paradox still remains: if you are a science fiction lover, perhaps you have noticed that a virtual reality world such as the one appearing in the *Matrix* movies also offers an obvious solution to the Fermi paradox. By the way, do you feel something on your neck...?

Appendix

Appendix A: Drake's Equation

It has been said sometimes that it is impossible to write a book about extraterrestrial life without making use of the Drake equation or without bringing it up in some way. Well, as you have seen, it is not impossible, but just in the case you missed it, here it is.

Drake's equation had its origin in the 1961 meeting organized by Frank Drake, often considered the first SETI congress. In order to have an agenda to follow during the meeting, Drake formulated an equation, summing up all the points he considered relevant for the search of extraterrestrial intelligences. Its result gives an estimate of the amount of civilizations in our galaxy, susceptible to having detectable radio emissions. That equation is the famous Drake equation:

$$N = R \cdot f_p \cdot n_e \cdot f_l \cdot f_i \cdot f_c \cdot L$$

where N is the number of detectable civilizations. The other factors are: R: the rate of stellar formation, f_p is the fraction of stars having planets, n_e: the average number of planets per planetary system that can potentially support life, f_l: the fraction of these planets that actually develop life, f_i: the fraction of inhabited planets that develop intelligence, f_c: the fraction of planets with intelligence able to establish interstellar communications, L: the mean time in which such civilizations are detectable (also call sometimes the contact window).

We have to realize that the Drake equation does not give an estimate of the number of civilizations in the galaxy, but of the number of civilizations with a technology that makes them detectable by means of radio waves or some similar medium. After all, this is what really matters to SETI. An advanced civilization without a technology that makes it noticeable will be impossible to detect through SETI, and thus will not count for these calculations. By giving appropriate values to these parameters, it is possible to calculate how many detectable civilizations there are at present in our galaxy.

The parameters are sorted from less speculative to more. Although nowadays we have good values for some of these parameters, such as star formation rate, many others remain completely unknown; thus the calculation of any value of the Drake equation is merely speculation.

When Drake evaluated his equation for the first time, the value of the parameters were even less known. The estimate he obtained was very high, about 10,000 civilizations in the galaxy. Nowadays, astronomers trying to evaluate this equation are less optimistic, and many estimates range from tens to hundreds of civilizations.

But we really do not know. As said by Barney Oliver, former director of HRMS, "Drake's equation is a wonderful way of compressing a large amount of ignorance into a small space." The real value of Drake's equation is historical: it became a point of intersection between the different disciplines, a place where different types of scientists could work together, and it has turned out to be a useful way to organize the work of many researchers in SETI and in astrobiology. In fact, it stands as one of the first indications that this great task – finding intelligent and technologically savvy civilizations – would need a multidisciplinary approach to be successful.

Appendix B: A Shape for the Intelligence

Are there intelligent beings on other planets? And if so, what form would they have?

Although we do not know the answer to both questions, science fiction has been not ashamed to answer "yes" to the first question, and even to hazard several answers to the second one, giving several possible shapes to their bodies. We all know rather well the appealing face of the first scientific officer of the starship *Enterprise*, Mr. Spock, hybrid of human and Vulcan. Just a couple of pointed ears and *voilà*, you are an alien! Of course, in this case, the lack of scientific rigor was mostly due to the low budget the original *Star Trek* series had. But bigger budgets did not advance change much in their designs, and the preference for the anthropomorphic shape is very commonplace. Both in science fiction movies and books, biped humanoids abound, having more or less varied shapes, from the classical big-headed green and short Martians to the very assorted range of humanoids in the *Star Wars* series of films. Unfortunately, except on very few occasions, we do not find in these works anything that reflects how a technologically intelligent species might evolve on other worlds. Maybe it is just a lack of imagination. Or maybe we humans can only believe that someone is really intelligent if it looks like us.

A curious exception are the Martians of *The War of the Worlds* novel by H.G. Wells. Wells was a Victorian author of what we might today call "hard science fiction," science fiction that tries to be strictly rigorous with the well documented science of its period. Wells was a connoisseur of the science of his period, and he knew very well Darwin's theory of evolution by natural selection. Thus he knew that it was almost impossible that an alien intelligence, with an evolutionary history completely different from ours, developing on a different planet, could have a humanoid aspect. Therefore, Wells's Martians were a kind of big central mass that included the brain, the mouth, and the organs of the senses. From this mass emerged tentacles that provided these beings with manipulative capabilities, something that seems to be absolutely essential for any intelligent civilization that develops technology.

Wells's nightmare cephalopods could be upright over their tentacles on their native planet, but on Earth, due to its bigger gravity, they had to walk dragging themselves along the ground. Of course, this clever recognition of Wells was "corrected" in both movie versions (in 1953 by Byron Haskin and more recently, in 2005, by Steven Spielberg), where Martians are turned into big-headed short humanoids (does this sound familiar?).

Of course, despite the efforts of Wells to be scientifically correct, it seems he could not avoid a certain amount of earthling chauvinism when he designed his Martians: he give them a characteristic that we share with many animals on Earth – bilateral symmetry. However, this is not an essential characteristic for living beings at all. Plants in general do not have bilateral symmetry, and neither have animals such as the jellyfish, the sea urchin, or the starfish.

One might think, then, that it was not necessary for Wells's Martians to have bilateral symmetry. But is that true? In this case was it really earthling chauvinism, or was it the intuition of the writer?

Why do we have bilateral symmetry? Why do other beings not have it? As Martin Gardner brilliantly tells in his book *The Ambidextrous Universe*, the origin of our symmetries and asymmetries is imposed from the outside. In general, the three spatial dimensions are equivalent, and there is a symmetry among them. If we were floating in interstellar space, there would be no way to decide which dimension is most important in respect to the other two. But we are not floating in space. We live in a world with a gravitational field that always points towards Earth's center. This defines for every living being on the planet what is "up" and what is "down." It breaks the symmetry of the three dimensions.

Now, the up-down dimension is special. Living beings have to live "fighting" against this gravitational field, and this produces in them an up-down asymmetry. This is true even in the case of aquatic beings, since objects that are denser than water sink down and those less dense float up. If we put a living being upside down, we will notice instantaneously the difference, no matter if it is a jellyfish, a tree, or a man! Your feet are different from your head. The exception is found in simple single cell organisms. In this case the viscosity of water at their size scale prevents them from floating or sinking. For all practical purposes, for them there is no up or down. That is why we find so many single-cell organisms that do not clearly show an upper or a lower part.

Thus, one should expect this up-down asymmetry in complex living beings on other planets, too. What about the ahead-behind asymmetry? This develops when the organism has the need to move forward quickly. Organisms that do not move, such as plants or corals, have only axial symmetry; there is no way to determine which is its front part. For them, there is no privileged horizontal direction. Something similar happens with organisms that move slowly, such as starfish. But if one has the need to move fast, to be efficient one has to break that symmetry; it is indispensable to have a front part and a back part. So when there is the ahead-behind asymmetry and the up-down asymmetry, the organisms have bilateral symmetry. And this is why an airplane will always be more efficient for flying in the skies than a flying saucer.

So most animals keep their bilateral symmetry. Of course, there are some exceptions, as for example when a fish decides to lie down on the bottom of the sea always on the same side, such as soles do, in a direction where there is sand and in the opposite direction there are predators. This forces the organism to break its bilateral asymmetry, giving soles their peculiar aspect.

Summing up, if extraterrestrial beings move in an efficient way, we should expect that they will have bilateral symmetry.

Another guide to guessing the shape of alien beings is convergent evolution. As we saw, dolphins, sharks, and ichthyosaurs share the same aerodynamic form, due to the fact that it is a very good solution for moving fast through water. A similar shape should be expected in alien creatures that are quick-swimming. Many convergent evolution examples are found on our planet, giving statistical support to some evolutionary solutions. For example, eyes evolved independently 40 times in different lineages on Earth. This is one of the best examples of convergent evolution. This fact should not be a surprise, as having eyes provides many advantages! Therefore, if light is an efficient way to transmit information in our alien's world, it would be rather probable that the aliens have eyes too; possibly an even number of eyes, thanks to bilateral symmetry. And, of course, they will be placed in a location that allows the organism to observe forwards, in the direction of its movement; in a frontal position for instance (although some additional rear eyes to take care of predators from behind would not be a bad idea). One should expect that alien beings will develop other senses that have also appeared many times on Earth through convergent evolution, as in the senses of touch, smell, and hearing (if sound waves are an efficient way to transmit information on our alien's world, of course).

Finally, if our aliens develop a technological civilization, they will certainly have some manipulation abilities, such as creating tools and using them. On Earth different solutions to manipulate the environment have been found – the trunk of elephants (a modified lip), the tongues of giraffes, hands (and not only in primates – keep in mind the raccoons, for instance), tentacles (in cephalopods, corals, and other organisms), protruding bones (as in the panda's thumb), jaws of ants and other insects, and so on. In short, there is a wide range of possibilities.

Appendix C: The Dimensions of Life

When thinking about the topic of life in the universe, sooner or later one arrives at this question: What are the minimum essential requirements for life to exist? Is water necessary? Is carbon chemistry indispensable? Should there be a star close to the planet? Are planets essential?

For many people, the correct answer to this question is the one that Carl Sagan gave in his famous apple pie recipe: "If you wish to make an apple pie from scratch, you must first invent the Universe." This is not obvious. Although it is something that we take for granted, life could not exist without three-dimensional space, where things exist, and time also exists and allows processes to occur. At least, both factors should exist. And both appeared with the universe.

One moment! We have said "three-dimensional" space. Why didn't we simply say "space?" It seems obvious that, without any kind of space, there could not be living beings, but why necessarily three-dimensional? After all, some organisms, such as flatworms, are practically flat. As they do not have a circulatory system, their cells must receive oxygen by direct diffusion through the skin, and their body must be extremely flat to assure all of the organism gets its share of oxygen. Flatworms seem to live perfectly well in two dimensions.

Something similar happens with cheela, the intelligent characters of the novel *Dragon's Egg* by Robert Forward. Dragon's Egg is the name of a neutron star, where nuclear reactions play an analogous role to chemical reactions in our world. There, a complete ecosystem of living beings made up of neutron matter has emerged who, due to the enormous surface gravity of their star (several billions of g) are extremely flat. Mountains in Dragon's Egg are only a millimeter high, but a fall from one of them is fatal.

Cheela doctors do not need special devices to study the innards of their patients. It is enough to stand up and see above the patient to see its interior.

These organisms are interesting examples that offer solutions to some of the problems of living in two dimensions. But "practically flat" is not the same as "flat," and although the third dimension seems not to have relevance in the daily life of cheela and flatworms, in fact, they could not live without it (flatworms receive their share of oxygen "from the top," for instance). Is life in a strictly two-dimensional world impossible, then?

The novel *Flatland*, published in 1884 by the English clergyman Edwin A. Abbott, presents a two-dimensional world inhabited by truly flat organisms. The main character, A. Square, is a square that has the opportunity to travel to the third dimension and to visit one-dimensional beings. But could this two-dimensional character have a digestive tract crossing his body?

Stephen Hawking, who thinks two-dimensional life is impossible, assures us that he could not, as the organism would be separated into two parts. But if the digestive tract did a strong zigzag, both parts could remain joined, as pieces in a jigsaw puzzle, although there would not be communication between the two parts. In this case, one could think that both parts are two symbiotic organisms cooperating, sharing a digestive tract. Another possibility could be an organism with a stomach bag with a unique opening that plays the double role of anus and mouth, as is the case of our three-dimensional hydras.

Nevertheless, with less than three dimensions, the complexity of life drops dramatically. In fact, the impediments for life in a two-dimensional world could be even greater, as the great majority of molecules composing organisms have a three-dimensional structure that is as important to them as is their chemical composition. Besides, even interactions among two-dimensional molecules (such as water) in many cases happen only through the third dimension.

And what happens with life in four or more dimensions? One might think that everything that happens in three dimensions would also happen in four, as the first is a subset of the second. And if the three-dimensional case is more complex than the two-dimensional case, with four dimensions we will have still more complexity. But...

If we believe in superstring theories, which try to unify quantum physics with Einstein's relativity, the universe has a minimum of nine spatial dimensions. (Some models postulate even 25 dimensions!) The problem is that we only see three. How could we reconcile observation with theory? By means of an astute trick.

To understand, let us examine the moment just before the Big Bang. If we look around we will see that the universe is extremely simple: there is no matter, no energy, only really tiny empty space, with a size billions of times smaller than an atom. But the enormous energy density this small empty space has is so big that it curves the space around it. If we were to travel this space in a straight line (in any direction of its numerous dimensions), we would return to the starting point just after traveling only 10^{-20} cm. The trip would be so short, in fact, that we would not notice that we had done it. And this is the key.

Just after the Big Bang was the inflationary epoch, an accelerated expansion of the universe caused (simplifying the story) by the sudden appearance of an incommensurable avalanche of Higgs bosons (particles predicted by quantum field theory). There appear to be so many in that minuscule original universe that, simply, there is not room for them. The immense pressure of so many particles in such a confined space produces the accelerated expansion of the space to make room. This is what is called the Big Bang.

But not all the dimensions experienced the Big Bang – only three! The remaining dimensions are as small as they were originally. So, if you travel in a straight line through one of these other dimensions (and possibly you are doing that at this very moment) you will return to the starting point only after traveling 10^{-20} cm; you will not notice it! This way, superstring scientists can solve the absolute discrepancy between their multidimensional models and observations. However, proponents of this theory do not offer any explanation about why three dimensions, and only three, had this fate.

So here is another possibility. It happens that three-dimensional space has an exclusive property that other dimensions do not share. This is a mathematical operation between vectors that can only happen in three dimensions – the cross product. It requires two vectors to produce a third vector perpendicular to the other two. In two-dimensional space, the third vector will be outside this space! And in four or more dimensions, perpendicularity to two vectors is not well defined. Only in three dimensions does the cross product make sense. And it happens that many fundamental laws of the universe are cross products, such as the magnetic field, the angular moment, the movement of electric charges – laws that will be senseless in non-three-dimensional space.

Did this force the existence of a universe with just three giant dimensions? Did this force life to be necessarily three-dimensional? Of course we do not know, but it makes us view boring lessons about vectors with new eyes. And that is something.

Appendix D: Bibliography

Jakosky, Bruce. *Search for Life on Other Planets.* Cambridge University Press, London, 1998.

Luque, Bartolo, and Fernando Ballesteros, et al. "Astrobiología, un puente entre el Big Bang y la Vida". Ed. Akal, 2009.

Oparin, Alexei. *Origin of Life.* Dover Publications, 1965.

Altschuler, Daniel R. *Children of the Stars.* Cambridge University Press, London, 2002.

Asimov, Isaac. *Robots and Empire.* Collins, 1994.

Lowell, Percival. *Mars as the Abode of Life.* Kessinger Publishing, 2008.

Hoyle, Fred. *Astronomical Origins of Life – Steps Towards Panspermia.* Springer, 1999.

Dick Steven J. *Life on Other Worlds: The 20th Century Extraterrestrial Life Debate.* Cambridge University Press, 2001.

Drake, Frank, and Dava Sobel. *Is Anyone Out There? The Search for Extraterrestrial Intelligence.* Pocket Books, 1997.

Drake, Frank, et al. *Carl Sagan's Universe.* Cambridge University Press, 1997.

Sagan, Carl, et al. *Communication with Extraterrestrial Intelligence* (summary of the 1971 CETI conference), The MIT Press, 1975.

Shklovskii, I, and Carl Sagan. *Intelligent Life in the Universe.* Emerson-Adams Press, 1998.

Sagan, Carl. *Contact.* Pocket Books, 1997.

Niven, Larry. *Ringworld.* Del Rey, New York, 1985.

Steffes, P.G. "Search for extraterrestrial intelligence/high resolution microwave survey team member final report for grant NAG, 2,700, report period, March 1, 1991 through August 31, 1994 (SuDoc NAS 1.26:196478)". Ed. NASA National Technical Information Service, 1994.

Shostak, Seth, and Frank Drake. *Sharing the Universe: Perspectives on Extraterrestrial Life.* Berkeley Hills Books, 1998.

Pinker, Steven. *The Language Instinct,* Harper Perennial Modern Classics, 2007.

"Sensory Abilities of Cetaceans: Laboratory and Field Evidence," NATO ASI Series. Series to: *Life Sciences* Vol. 196, 1991.

Dewey, John. *How We Think.* Dover Publications, 1997.

Dos Santos, Marcelo. "El Manuscrito Voynich," Ed. Aguilar, 2005.

Jacobs, Lou. *By Jupiter!: The Remarkable Journey of Pioneer 10.* Hawthorn Books, 1975.

Sagan, Carl. *The Cosmic Connection.* Dell, 1975.

Sagan, Carl. *Murmurs of Earth: The Voyager Interstellar Record.* Ballantine Books, 1979.

Freudenthal, Hans. "Lincos: Design of a Language for Cosmic Intercourse, Part I." Ed. *Journal of Symbolic Logic,* Volume 38, Issue 3 (1973), 517.

Webb, Stephen. *If the Universe Is Teeming with Aliens... Where Is Everybody? Fifty Solutions to Fermi's Paradox and the Problem of Extraterrestrial Life.* Springer, 2002.

Lem, Stanislav. *His Master's Voice.* Northwestern University Press, 1989.

Walter, van Beek. "Dogon Reestudied: A Field Evaluation of the Work of Marcel Griaule." *Current Anthropology,* Volume 32, Issue 2, 139–167.

Index